LABORATORY GUIDE
FOR HUMAN ANATOMY

LABORATORY GUIDE FOR HUMAN ANATOMY

William J. Radke, Ph.D.
The University of Central Oklahoma

JOHN WILEY & SONS, INC.

ACQUISITIONS EDITOR	Bonnie Roesch
ASSOCIATE EDITOR	Mary O'Sullivan
MARKETING MANAGER	Clay Stone
SENIOR PRODUCTION EDITOR	Norine M. Pigliucci
SENIOR DESIGNER	Karin Kincheloe
ILLUSTRATION EDITOR	Anna Melhorn
PHOTO EDITOR	Hilary Newman
TEXT DESIGN	Lee Goldstein
COVER DESIGN	Carol Grobe
PRODUCTION MANAGEMENT SERVICES	TechBooks

This book was set in 10.5/13 Times Roman by TechBooks and printed and bound by Courier/Westford. The cover was printed by Phoenix Color.

This book is printed on acid-free paper. ∞

Library of Congress Cataloging-in-Publication Data

Radke, William J.
    A laboratory guide to human anatomy / William J. Radke.
            p. cm.
    Includes bibliographical references and index.
    ISBN 0-471-41413-1 (paper : alk. paper)
    1. Human anatomy—Laboratory manuals.   I. Title.

QM34 .R3176 2001
611—dc21   2001044409

Printed in the United States of America

10 9 8 7 6 5 4 3 2 1

To Bob Chiasson who made it possible.

To Chris, Sarah, and Julia who make it worthwhile.

Read This First

This manual is for the student and the instructor. It is designed as a teaching tool, not as a reference. The anatomy/physiology student will use the manual as a guide to the study of the human body. The instructor may use the manual in two ways: as a source of overhead transparency material for lectures involving human systems and as a laboratory manual for students. This manual may be used in a self-guided instructional approach or as part of a traditional laboratory-lecture format.

USER'S GUIDE

Proper Labeling

You should *label* all the figures in this manual. Labels should be placed to the left and right of the drawings and lines ruled from the label to the structure. These lines should not cross. Labels should be printed. Terms to be applied as labels are found on the same page as, or the page facing the diagram. No learning is accomplished by using letters, numbers, or colors to label diagrams. Be sure to **write out each term on the diagram.** *Use* pencil not pen.

Coloring

You may find it helpful to shade some drawings with colored pencils. This aids in separating adjacent structures but should be done sparingly. It is not necessary for each structure to be colored differently because labels are used for identification. This manual is not intended to be a coloring book. Coloring is time consuming and is not a useful learning tool.

Glossary

A glossary of terms is found at the end of this manual. The glossary contains terminology that appears, but is not defined, in the text.

Color Plates

At the end of this manual is a Histology Atlas and a Cat Dissection Atlas. These provide photographs of tissues and cat dissections for your reference.

Other Aids

Words that appear in **bold print** are significant terms that may not appear in the lists associated with the diagrams. Terms that appear in *italics* provide instruction. When dissecting, these terms will provide the information you need to complete the exercise—for example, "*reflect* the trapezius muscle." This same designation is used in the labeling exercises to provide directions.

⚠ THERE IS POTENTIAL FOR PERSONAL INJURY OR RISK OF DAMAGE TO LABORATORY EQUIPMENT OR SPECIMENS. BE *VERY CAREFUL* WHEN YOU SEE THIS SYMBOL.

Laboratory Safety Procedures

- *Keep* your work area clear of clothing, backpacks, and other items not related to the laboratory activity.
- *Avoid* contact with embalming fluids. Wear gloves throughout the laboratory period.
- *Wear* safety glasses.
- *Ask* your instructor how to properly mount and remove a scalpel blade if these are to be used in your laboratory.
- *Do not* eat or drink in the laboratory.
- *Keep* your hands and instruments away from your face.
- *Return* all equipment, models, specimens, slides, and other items to the proper storage area at the end of the laboratory period.

- *Wipe* down your laboratory table, instruments, and dissection tray when you finish.
- *Wash* your hands when you are through in the laboratory.

Textbook Reference

This manual may be used in laboratories that have or do not have access to cadavers. The labeling of diagrams is to be completed using a textbook of anatomy as a reference. *Ask* your instructor which textbook is preferred or *see* your course outline.

Pronunciation Guide

Pronunciation guides are provided for all terminology used in this manual. Each term is spelled phonetically immediately after it first appears in the text. The following general rules were applied in creating the guide:

1. The syllable with the strongest accent is capitalized: ah-NAT-o-me.
2. Vowels to be pronounced in the long form are underlined: blade, bite.
3. Unmarked vowels are pronounced in the short form: mitt, drum.
4. Other indicators of sounds are given: oo as in blue; yoo as in cute; oy as in foil.

Learning the correct spelling of a term is easier if you spend just a few moments studying the proper pronunciation. Frequently repeating the terminology found in this manual out loud will help commit the terms to memory.

Models and Dissection

For three-dimensional understanding and interpretation of the two-dimensional drawings in this manual, it is important that the laboratory be supplied with a selection of anatomical models, organ dissections, and either cats or human cadavers for dissection. It is recommended that the following be available:

Models and skeletal supplies
skull (real bone if possible)
whole skeleton (real bone if possible)
torso, either life-size or accurate reduced scale
peripheral nervous system model
superior appendage model (dissectable)
inferior appendage model (dissectable)
brain model (dissectable, midsagittal plane at least)
spine model (showing spinal nerves and sympathetic trunk)
heart model (dissectable)
fetal circulation model
bronchotracheal model
urogenital model (male and female)
kidney model (dissectable)
Preserved materials for display or dissection
sheep brain
sheep or pig heart
sheep or pig kidney
sheep or beef eye
cat
Prepared slides
simple squamous, cuboidal, and columnar epithelium
stratified squamous, cuboidal, and columnar epithelium
ciliated epithelium
areolar connective
adipose
reticular connective
dense regular connective
dense irregular connective
elastic connective
hyaline cartilage
fibrocartilage
elastic cartilage
hairy skin
glandular skin
Meissner's corpuscles
pacinian corpuscles
ground bone
smooth, cardiac, and skeletal muscle
blood (Wright's stain)
motor end plates
Charts
cardiovascular system
lymphatic system
nervous system
skeletal muscle system (anterior, posterior views superficial and deep)

The materials listed above should be available in the laboratory. It is difficult to develop spatial concepts in anatomy without them. You are encouraged to *manipulate* the specimens and models whenever possible. All anatomical structures illustrated in the diagrams of this manual should be *located* on the models and/or preserved materials. Other optional materials include the following:

Prosected cadavers (cadavers dissected in an advanced anatomy course work particularly well)

Human materials (from supply houses or state anatomical boards)

heart

whole brain

brain sections

placentae

kidney

fetus

disarticulated skull

X-rays (often available from local clinics with identification removed for privacy)

2 × 2 slides or videotape of cadaver dissection

Interactive computer software such as AnatLab by Elephant Software

Palpation

You may neglect one of the most available and perfect teaching aids in the laboratory—your own body. Most skeletal elements may be felt through the surface of the skin. Many features such as processes, spines, epicondyles, foramina, fossae, and tuberosities may be palpated. It is also possible to note many superficial veins and a few arteries. The special senses and oral cavity can be thoroughly explored in the laboratory. Tendons, ligaments, and muscles are very easy to locate on one's own body. Chapter 21 demonstrates topography and is best studied after completing the skeletal and muscle units. In this manual you will find an asterisk (*) preceding a structure that you are asked to locate on your own body. For example, "*olecranon process" instructs you to locate the olecranon process on your elbow.

Latin and Greek Roots

A dictionary of root words and combining forms or a medical dictionary should be available at all times during the laboratory. Knowing the meaning of the Latin or Greek word from which the term originates helps you to remember the term.

Critical Thinking

You will find questions to stimulate your critical thinking skills throughout this manual. These exercises are designed to help you evaluate your knowledge of the subject. You should not proceed to the next section or to an examination without attempting the problems.

The problems will evaluate your understanding of:

• Anatomical terminology

• Function (physiology)

• Integration (relationships between systems)

Don't consider this manual your only resource for a course in human anatomy. You must *read* your textbook and *use* other references as well. Your instructor may suggest such references. Your college or university library will have books on anatomy and physiology and carry journals (periodicals) from which you may obtain additional and supportive information. You should also *make use* of resources in your computer laboratory and those on the Internet.

Laboratory Reports complete most chapters. Your instructor may require that the reports be turned in. If not, you should *complete* the reports as part of your regular study. Since the reports may require information beyond that found in this manual, have a textbook of human anatomy available.

If your textbook contains a study outline for each chapter, *use* this as a guide to organize your study. Your textbook may also include sections regarding clinical applications of the information you have just covered. *Review* them carefully; they will help you integrate the anatomy and physiology of a system and will point out the relationships between systems. Finally, *do not overlook* whatever review questions are found in your textbook.

Contents

1 Introduction to the Human Organism

Contents

1

Objectives

1. Define and provide an example of each level of organization of life.
2. Define the term anatomical position and provide the reason for its use.
3. Define and use directional terminology as it is applied to the human body.
4. Describe the anatomical planes and describe the sections and views associated with each.
5. Define and delineate the regions of the human body.
6. List and locate the body cavities and the organs each contains.
7. List and locate the abdominopelvic regions and the organs found in each.

A DEFINITION OF ANATOMY

Anatomy (ah-NAT-o-me), meaning "to cut," is the study of structure. An understanding of anatomy is best obtained by knowing the function of the parts. Therefore, as you study anatomy, always determine what each organ or organ system does.

ORGANIZATION OF LIFE

Before beginning the study of human anatomy, it is important to know how living things are organized. In your study, you will be concerned primarily with tissue, organ, and system levels of organization. The following table (see p. 2) describes one way in which life may be organized.

1

Level	Component	Description
Chemical Level	**Subatomic** (ah-TOM-ik) **particles**	Atoms are formed from these tiny particles. They include protons (PRO̱-tons), neutrons (NYOO-trons), and electrons (e̱-LEK-trons).
	Atom	All matter is made of units called atoms. Each atom has a unique combination of subatomic particles and, as a result, has different physical properties.
	Molecule (MAHL-e̱-kyool)	The molecule is a unique combination of atoms. In living things the most abundant organic molecules are carbohydrates, lipids, proteins, and nucleic acids.
Cellular Level	**Organelle** (oṟ-gan-EL)	Cells are made up of specialized structures called organelles. These are complex arrangements of molecules that carry out the functions of the cells.
	Cell (sel)	The cell is the basic functional and structural unit of an organism.
Tissue Level	**Tissue**	Tissues are groups of cells of a similar structure, function, and embryological origin. The four tissue classes are (1) epithelial (ep-i-THE̱-le̱-al), (2) connective, (3) muscle, and (4) nervous.
Organ Level	**Organ**	An organ carries out a specific function and consists of two or more different kinds of tissues. The heart is an example of an organ.
System Level	**System**	Two or more organs carrying out a common function form an organ system. For example, the heart, vessels, lymphatics, and blood form the cardiovascular system.
Organism Level	**Organism** (OṞ-gah-nizm)	A single individual.
Population Level	**Population**	Numerous organisms of the same species form a population. For example, all of the humans taking this human anatomy class is a population.
Community Level	**Community**	In a given location it is unlikely that there will be only one kind of organism present. In your anatomy laboratory, there will be humans, many species of bacteria and fungi, as well as some spiders and perhaps a cockroach or two. This combination of interacting species is a community.
Ecosystem Level	**Ecosystem** (E-ko̱-sis-tem)	All the characteristics of a geographic area: climate, abiotic, and biotic.

CRITICAL THINKING

Practice your understanding of the organization of life by completing the following.

1. Human anatomy considers which three levels of organization?

   _____

   _____

   _____

   _____

2. Injury to the body at cellular, tissue, or organ levels affects function at the _____

   _____ level.

3. The integument is an example of an _____

   _____.

4. The kidney is an example of an _____

   _____.

5. A blood vessel is lined by a _____

_____.

ANATOMICAL POSITION

Anatomical position implies that a subject

is standing upright

has the head level

has the feet flat on the floor with toes straight ahead

has the arms at the side with the palms turned forward.

The purpose is to create a convention in which illustrations, photographs, and directional terminology may be assumed to be consistent (see Figure 1.1).

CRITICAL THINKING

Complete the following:

How might an illustration of the forearm be confusing if the appendage is not in anatomical position?

_____

_____

_____

_____

DIRECTIONAL TERMINOLOGY

Before beginning with the detailed study of human anatomy, you should *remember* the following directional terminology. You will use these terms throughout your study of anatomy and will continue to use them in your chosen health-related field.

Using your textbook as a reference, *define* the following terms.

superior (soo-PE-re-or) _____

_____

inferior (in-FE-re-or) _____

_____

anterior (an-TE-re-or) _____

_____

posterior (pos-TE-re-or)_____

_____

medial (ME-de-al) _____

_____

lateral (LAT-er-al) _____

_____

proximal (PROK-sih-mal) _____

_____

distal (DIS-tl)_____

_____

superficial (soo-per-FISH-al) _____

_____

deep (DEP) _____

_____

CRITICAL THINKING

Use the terms you have just learned to *describe* the following. You may have to use your textbook to look up some of the anatomical structures.

1. The epidermis is _____

_____ to the dermis.

2. The spine is _____

_____ to the sternum.

3. The chest is _____

_____ to the abdomen.

4. The hand is _____

_____ to the forearm.

5. The sternum is _____

_____ to the heart.

6. The axillary border of the scapula is _____

_____ to the vertebral border.

7. The hips are _____

_____ to the shoulders.

8. The knee is _____

_____ to the ankle.

9. The cranium is _____

_____ to the scalp.

10. In anatomical position the thumb is _____

_____ to the ring finger.

PLANES, SECTIONS, AND VIEWS

Using your textbook as a reference, *label* the planes illustrated in Figure 1.1. As you do so, *indicate* the name of the section created when the body is cut along the plane. Finally, *name* the views made possible by the section.

midsagittal (mid-SAJ-ih-tal) or median plane

parasagittal (par-ah-SAJ-ih-tal) plane

frontal (FRUN-tal) plane

transverse (trans-VERS) or horizontal plane

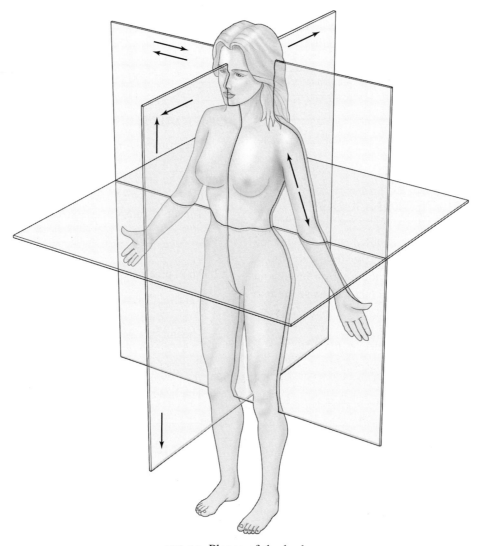

FIG. 1.1 Planes of the body.

CRITICAL THINKING

Use the terms you have just learned to *name* the following. You may have to *use* your textbook to look up some of the anatomical structures.

1. *Name* the section in which both the ethmoid and temporal bones may be seen.

   _____

2. *Name* the section in which the entire width of the sternum may be seen.

   _____

3. *Name* the plane that divides the body into mirror image halves.

   _____

4. A section that divides the forearm into proximal and distal portions is a _____ section.

5. A section parallel to the midsagittal section is a _____ section.

6. A section removing just the tip of the nose parallel to the face is a _____ section.

7. Looking at the cut surface of the removed nose is a _____ view.

8. Looking at the bottom of a brain removed from the cranium is a _____ view.

9. Looking at the heart from the right side is a right _____ view.

10. *Assume* a midsagittal section of the brain. When looking at the cut surface, this is a _____ view.

REGIONAL TERMINOLOGY

Locate the following regions and *label* parts (a) and (b) in Figure 1.2.

head

    cranium (KR<u>A</u>-n<u>e</u>-um)

    face

neck

trunk

    back

    thorax (TH<u>O</u>-raks)

    abdomen (ab-D<u>O</u>-men)

    pelvis (PEL-vis)

appendages (extremities)

    superior appendage (ah-PEN-dij)

        axilla (ak-SIL-ah)

        shoulder

        brachium (BR<u>A</u>-k<u>e</u>-um) or arm

        elbow

        antebrachium or forearm

        carpus (KAR-pus) or wrist

        manus (MA-nus) or hand

           palm

           dorsum (D<u>O</u>R-sum)

           pollex (P<u>O</u>L-eks) or thumb

           digit (DIJ-it)

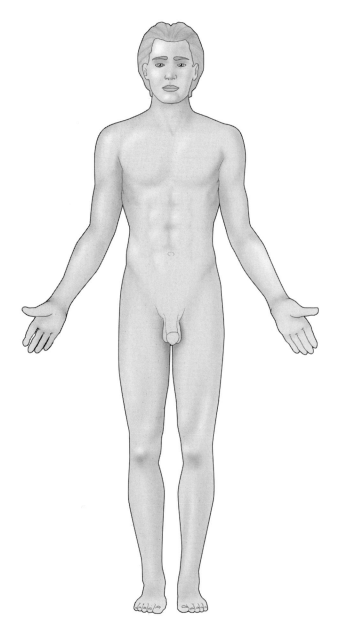 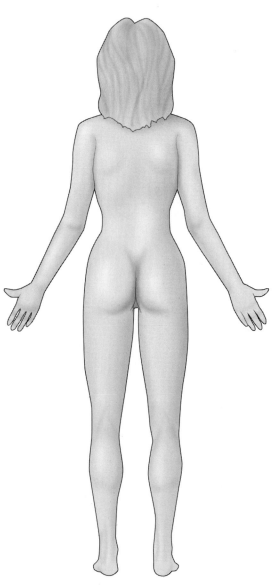

FIG. 1.2 Anatomical position. (a) Male, anterior view. (b) Female, posterior view.

inferior appendage

    buttocks

    thigh

    knee

    leg

      calf

    tarsus (TAR-sus) or ankle

    pes or foot

      heel

      sole

      dorsum

      hallux (HAL-uks) or great toe

      digit

CRITICAL THINKING

Use regional terminology to *complete* the following. *Use* your textbook or a medical dictionary to *look up* unfamiliar terms.

1. The bony structure containing the brain is the _____.

2. An English word derived from the Latin for hand is _____.

3. Axillary hair is located in the _____ _____.

4. The posterior side of the hand when in anatomical position is the _____ _____.

5. The anterior trunk, from superior to inferior, consists of the _____, _____, and the _____.

6. The region consisting of seven cervical vertebrae is the _____.

7. The segment proximal to the antebrachium is the _____.

8. The thigh and leg articulate at the _____.

9. Digit number one of the foot is the _____.

10. A cubit is the distance from the tip of the middle digit to the _____.

ABDOMINOPELVIC REGIONS

In Figure 1.3, *locate* and *label* the following abdominopelvic regions and *list* the visceral organs found in each.

right hypochondriac (hi-po-KON-dre-ak)

left hypochondriac

epigastric (ep-e-GAS-trik)

right lumbar (LUM-bar)

umbilical (um-BIL-ih-kle)

right iliac (IL-e-ak)

left iliac

hypogastric (hi-po-GAS-trik)

CRITICAL THINKING

1. In which abdominopelvic region is a surgical approach to the vermiform appendix made?

   _____

2. What obvious superficial feature is located in the umbilical region?

   _____

3. The _____

   region contains the urinary bladder.

4. The majority of the stomach is located in the

   _____ region.

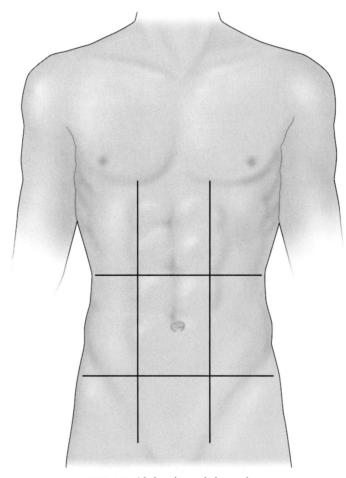

FIG. 1.3 Abdominopelvic regions.

5. The _____

region contains the ascending colon.

BODY CAVITIES

Locate and *label* the following body cavities in Figure 1.4. *List* the primary organ(s) found in each cavity.

cranial (KRĀ-nē-al) cavity

vertebral (VER-teh-bral) cavity

pericardial (per-ē-KAR-dē-al) cavity

pleural (PLOO-ral) cavity

abdominal cavity

pelvic (PEL-vik) cavity

CRITICAL THINKING

Complete the following:

1. The _____

cavities contain the lungs.

2. A transverse section through the umbilicus will

pass through the _____

cavity and the _____ cavity.

3. The _____ cavity

and _____

cavity contain the central nervous system.

4. List the subdivisions of the ventral and dorsal body cavities.

_____

_____

_____

_____

5. The abdominal cavity is continuous with the

_____ cavity.

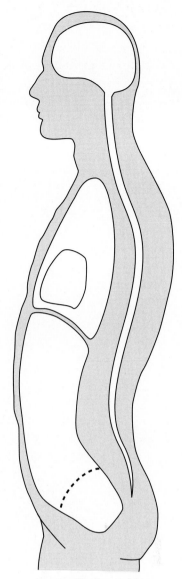

FIG. 1.4 Body cavities.

SHORT ANSWER

1. *Describe* the following regions:

cranial

orbital

nasal

buccal

thoracic

pectoral

abdominal

axillary

brachial

cubital

antebrachial

pelvic

inguinal

pubic

gluteal

femoral

popliteal

tibial

plantar

MATCHING

Match the descriptions in column A with the terms in column B.

<table>
<tr><td colspan="2" align="center">**A**</td><td align="center">**B**</td></tr>
<tr><td>_____</td><td>1. Proximal region of the superior appendage.</td><td>a. anterior</td></tr>
<tr><td>_____</td><td>2. Divides the body into unequal left and right halves.</td><td>b. brachium</td></tr>
<tr><td>_____</td><td>3. The dermis is _____ to the epidermis.</td><td>c. distal</td></tr>
<tr><td>_____</td><td>4. The left hypochondriac region is _____ to the left lumbar region.</td><td>d. deep</td></tr>
<tr><td></td><td></td><td>e. hallux</td></tr>
<tr><td>_____</td><td>5. The popliteal fossa is _____ to the knee cap.</td><td>f. inferior</td></tr>
<tr><td>_____</td><td>6. The first toe.</td><td>g. midsagittal section</td></tr>
<tr><td>_____</td><td>7. The pes is _____ to the leg.</td><td>h. parasagittal section</td></tr>
<tr><td>_____</td><td>8. The neck is _____ to the cranium.</td><td>i. pericardial cavity</td></tr>
<tr><td>_____</td><td>9. The palm is _____ to the dorsum of the hand.</td><td>j. posterior</td></tr>
<tr><td>_____</td><td>10. A section showing vertebrae, ribs, and sternum.</td><td>k. proximal</td></tr>
<tr><td>_____</td><td>11. A section in which the entire vertebral column may be seen.</td><td>l. superior</td></tr>
<tr><td>_____</td><td>12. Location of the spinal cord.</td><td>m. transverse section</td></tr>
<tr><td>_____</td><td>13. Location of the heart.</td><td>n. vertebral cavity</td></tr>
<tr><td>_____</td><td>14. The cubital region as compared with the carpus.</td><td></td></tr>
</table>

Match the organ in column A with the body cavity in which it is located in column B.

A	B

A

_____ 1. spinal cord

_____ 2. uterus

_____ 3. brain

_____ 4. vermiform appendix

_____ 5. heart

_____ 6. lung

B

a. abdominal

b. cranial

c. pelvic

d. pericardial

e. pleural

f. vertebral

MULTIPLE CHOICE

_____ 1. In the organization of life, which of the following group to form organ systems?

    a. cells   c. molecules

    b. organelles   d. organs

_____ 2. Which of the following group to form tissues?

    a. cells   c. molecules

    b. organelles   d. organs

2 Cells

Contents

Objectives

1. Name and define the parts of the compound microscope.

2. Demonstrate the correct use of the compound microscope on scanning, low, and high dry power.

3. Define a cell.

4. Describe the structure and function of the plasma membrane.

5. List and provide the function for the cell organelles.

6. Identify the parts of your own cheek cells.

THE COMPOUND MICROSCOPE

Because of the fragility and expense of the compound microscope, it is important to develop good habits of use and care. To begin, *do* the following:

 ALWAYS TRANSPORT YOUR MICROSCOPE BY PLACING ONE HAND UNDER THE BASE AND GRASPING THE ARM OF THE SCOPE WITH THE OTHER HAND.

PLACE THE MICROSCOPE ON YOUR LABORATORY TABLE AND *CONNECT* THE POWER PLUG TO THE OUTLET.

Refer to Figure 2.1 and *locate* the following on your microscope:

ocular lens (OK-u̱-lar lenz)

objective lenses (ob-JEK-tiv)

illuminator (ih-LOO-mih-na̱-to̱r)

illuminator switch

stage (sta̱j)

slide clips

coarse adjustment

fine adjustment

diaphragm (DI̱-ah-fram)

Using the Microscope

To view a microscope slide, *follow* these instructions.

1. *Using* the coarse adjustment, *increase* the distance between the stage and the objective lenses to the maximum.

2. *Rotate* the objective lens until the scanning objective (4×) snaps into place over the opening in the stage.

3. *Place* a slide on the stage, *clip* it down with the stage clips, and *center* the portion to be viewed over the opening in the stage.

4. *Using* the coarse adjustment, *bring* the objective as close as possible to the slide. It is not possible for the scanning objective to touch the slide, but it is good practice to *watch* how close the lens is to the slide.

5. Now, while *looking* into the ocular lens, slowly *rotate* the coarse adjustment to *move* the lens away from the slide until an image is observed.

6. *Sharpen* the image using the fine adjustment. If there seems to be an inappropriate amount of light, *adjust* the diaphragm to produce the best image. On scanning power, the diaphragm should be nearly closed. The image that you see is magnified 40 times, that is, 4× objective × 10× ocular = 40×.

Practice viewing with both eyes open to avoid eyestrain from squinting and also to avoid seeing your own eyelash in the field of view. One way to do this is to *leave* both eyes open but to *cover* the unused eye with your hand.

7. To *increase* the magnification, *turn* to the 10× objective (low dry power). Because your microscope is **parfocal** (par-FO̱-kal) you need only to *use* the fine adjustment to sharpen the image (10× objective × 10× ocular = 100×).

8. Finally, *rotate* the high dry objective (45×) into position and again *focus* with the fine adjustment only.

⚠ USE OF THE COARSE ADJUSTMENT WITH HIGH POWER MAY FORCE THE LENS INTO THE COVERSLIP OF THE SLIDE AND DAMAGE BOTH THE LENS AND THE SLIDE.

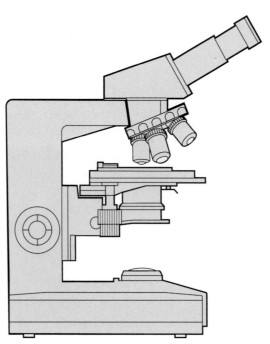

FIG. 2.1 Compound microscope.

9. The magnification here is 45× objective × 10× ocular = 450×. You will have to *increase* the amount of light by *opening* the diaphragm.

If you cannot see an image, *do* the following:

1. *Check* the light source.
2. *Check* the diaphragm.
3. *Make sure* the objective lens is properly snapped into position.
4. *Clean* the lens (⚠ With lens tissue only).
5. *Consult* your instructor.

When you finish, *do* the following:

1. *Return* to scanning power.
2. *Increase* the distance between the lens and slide to the maximum.
3. *Remove* the slide from the stage.
4. *Store* the power cord as instructed.
5. *Return* the microscope to storage.

Practice with the Microscope

Make a wet mount of newsprint by *following* these instructions:

1. *Cut* a letter out of the newsprint provided.
2. *Place* the letter on a microscope slide.
3. *Place* a drop of distilled water on the newsprint.
4. *Place* a coverslip over the newsprint and water.

CRITICAL THINKING

View the newsprint preparation and answer the following:

1. When you *move* the slide on the stage away from you, in what direction does the image move? Now, what happens when you *move* the slide

   toward you? _____

   _____

   to the left? _____

   _____

   to the right? _____

   _____

2. *Open* the diaphragm all the way on each power. What happens to the quality of the image in each case?

   _____

   _____

   _____

   _____

3. *Close* the diaphragm all the way on each power. What happens to the quality of the image in each case?

   _____

   _____

   _____

   _____

4. Once the newsprint is in perfect focus under each power, use the fine adjustment to *move* the objective lens slightly away from the slide. *Describe* what you see. What do you see when you *move* the lens a bit downward?

   _____

   _____

   _____

   _____

5. Note the position of the newsprint on the slide. When you *view* the newsprint through the microscope, is the image inverted or backward?

   _____

   _____

   _____

   _____

6. Practice moving the slide. Why is it always best to begin with scanning power when attempting to locate an object on the slide?

   _____

   _____

   _____

   _____

CELL STRUCTURE

Cytology (si-TOL-o-ge) is the study of cell biology (*cyte* = cell, *logus* = to study). In this section we will consider briefly the anatomy and function of the parts of the cell. You should *refer* to your textbook for the description and location of each of the following cell **organelles** (or-gan-ELZ).

On the opposite page, *draw* a typical cell, and *label* it with the terms below. Refer to HA1 in the Histology Atlas. Make sure that you *include* a description of the functions as you do so.

plasma (PLAZ-ma) membrane

cytoplasm (CI-to-plaz-im)

organelles

    nucleus (NOO-cle-us)

    endoplasmic reticulum (en-do-PLAZ-mik re-TIK-u-lum)

    ribosome (RI-bo-zom)

    Golgi complex (GOL-je)

mitochondria (mi-to-KON-dre-ah)

lysosome (LI-so-som)

microtubules (mi-cro-TOO-bul)

centrosome (SEN-tro-som)

vacuole (VAK-u-ol)

MICROSCOPE STUDY

1. Using the blunt end of a toothpick, gently *scrape* the inside wall of your cheek.
2. *Spread* the sample evenly across a clean microscope slide.
3. ⚠ *Dispose* of the toothpick in biohazard waste.
4. *Place* one drop of potassium iodide solution on the sample.
5. *Cover* with a coverslip.
6. *View* on scanning, low, and high dry power.
7. *Identify* the plasma membranes, cytoplasm, and nuclei of the cells.
8. ⚠ *Dispose* of the slide in broken glass waste.

This page is for your drawing of the cell.

SHORT ANSWER

1. *Define* the following:

   semipermeable (sem-e-PER-me-ah-bl)

   diffusion (di-FU-shun)

   osmotic (oz-MOT-ik) pressure

   active transport

   isotonic, hypotonic, and hypertonic (TAH-nik)

   phagocytosis (fag-o-si-TO-sis)

secretion

flagella (flah-JEL-ah)

cilia (SIL-e̱-ah)

mitosis (mi̱-TO̱-sis)

2. *Name* the cell organelles involved in synthesis and secretion of insulin.

3. *Describe* the events of mitosis and cytokinesis. What are the major steps?

4. Would you expect muscle, bone, or skin to have the most mitochondria per cell? Why?

5. Why are so few of the cell organelles visible in the cheek cell preparation?

Contents

3

Objectives

1. Define histology.

2. Define tissue.

3. Name the four tissue classes and describe each.

4. Name the germ layers of the embryo from which each class of tissue is derived.

5. Name and describe the different shapes and arrangements of cells in epithelial tissues.

6. Describe the structure, locations, and general functions for each of the epithelial tissue types.

7. Describe the structure, locations, and general functions for each of the connective tissues except bone and blood.

8. Identify each of the epithelial and connective tissues under the microscope and point out distinguishing characteristics of each.

TISSUE CLASSES AND ORIGINS

The study of tissues is **histology** (his-TOL-o-ge). The four **classes** of tissues comprising the human body are

1. **epithelial** (ep-ih-THE-le-al)

2. **connective**

3. **muscle**

4. **nervous**

In the embryo these tissues are formed from the

1. **Ectoderm** (EK-to-derm)—epithelial and nervous

2. **Endoderm** (EN-do-derm)—epithelial

3. **Mesoderm** (MES-o-derm)—connective and muscle

From the information provided in your textbook, *define* the function and *describe* the basic structure of each class of tissue.

epithelial _____

_____

connective _____

_____

muscle _____

_____

nervous _____

_____

EPITHELIAL TISSUES

Any description of epithelial tissues must contain a description of both the **arrangement** and the **shape** of the cells making up the tissue.

There are two basic arrangements. *Describe* the arrangements.

simple _____

_____

stratified (STRAT-ih-fīd) _____

_____

There are three basic shapes. *Describe* the shapes.

squamous (SKWA-mus) _____

_____

cuboidal (kyoo-BOI-dal) _____

_____

columnar (kō-LUM-nar) _____

_____

Label the epithelial tissues in Figures 3.1 and 3.2 (for example, part (a) in Figure 3.1 is simple squamous epithelial tissue). Where the tissue of interest may be unclear it is indicated with an arrow. Magnifications are 400× unless otherwise indicated (400× means the tissue in the figure is illustrated 400× larger than it really is and that high power should be the objective of choice for viewing). See Figures HA-2–8 for more information. *Supply* at least one location in the body where the tissue is found (the information is in your textbook).

simple squamous _____

_____

simple cuboidal _____

_____

simple columnar _____

_____

stratified squamous _____

_____

stratified cuboidal _____

_____

stratified columnar _____

_____

Some of these tissues have surface modifications (Figure 3.1b and HA-5) such as cilia, flagella, or microvilli. *Name* and *illustrate* a few of these tissues and *describe* their locations and functions below.

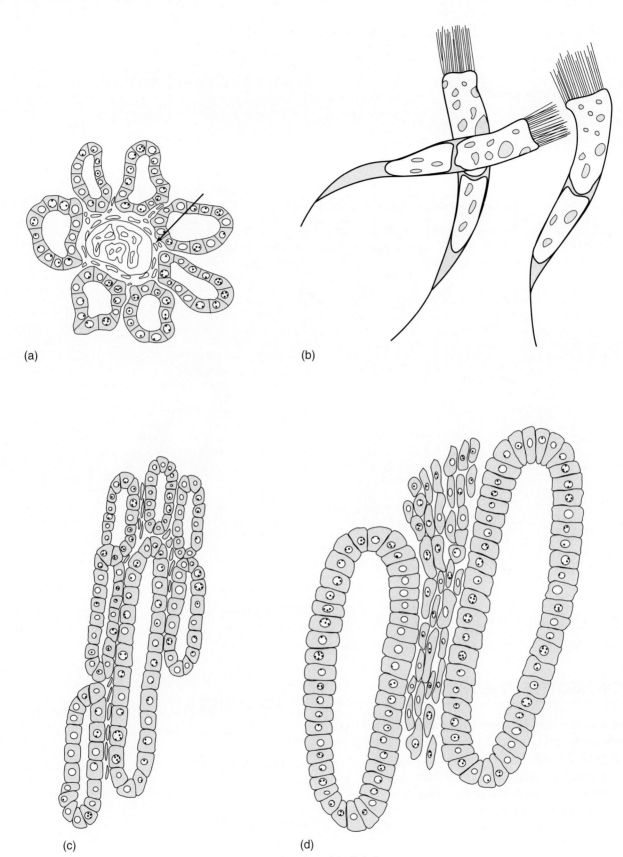

(a)

(b)

(c)

(d)

FIG. 3.1 Simple epithelial tissues.

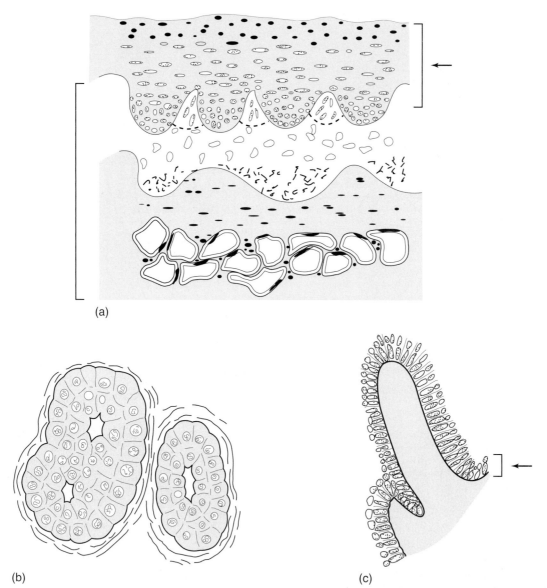

(a)

(b)   (c)

FIG. 3.2 Stratified epithelial tissues.

CONNECTIVE TISSUES

Connective tissues are more numerous and varied than epithelial tissues. *Label* the connective tissues found in Figures 3.3 and 3.4. Magnifications are 400× unless otherwise indicated. Where there may be confusion, an arrow points to the tissue. See Figures HA-9–17 for more information. After each tissue type shown below, *list* several locations in the body where the tissue is found.

loose connective tissues

    areolar (ah-RE-o-lar) _____

_____

    adipose (AD-ih-pos) _____

_____

    reticular (reh-TIK-u-lar)_____

_____

dense connective tissues

    dense regular _____

_____

    dense irregular_____

_____

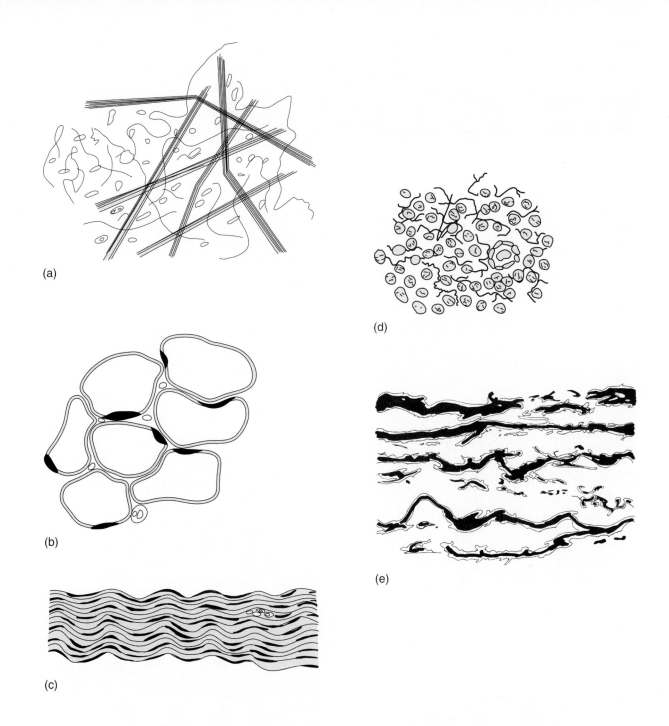

(a)

(b)

(c)

(d)

(e)

FIG. 3.3 Connective tissues.

elastic _____

_____

cartilage (KAR-tih-lij)

    hyaline (HĪ-ah-lin) _____

_____

    fibrocartilage _____

_____

    elastic_____

_____

osseous (OS-ē-us)

    bone (see Chapter 5, which discusses the skeletal
    system) _____

_____

vascular (VAS-kyoo-lar)

    blood (see Chapter 12, which discusses the car-
    diovascular system)_____

_____

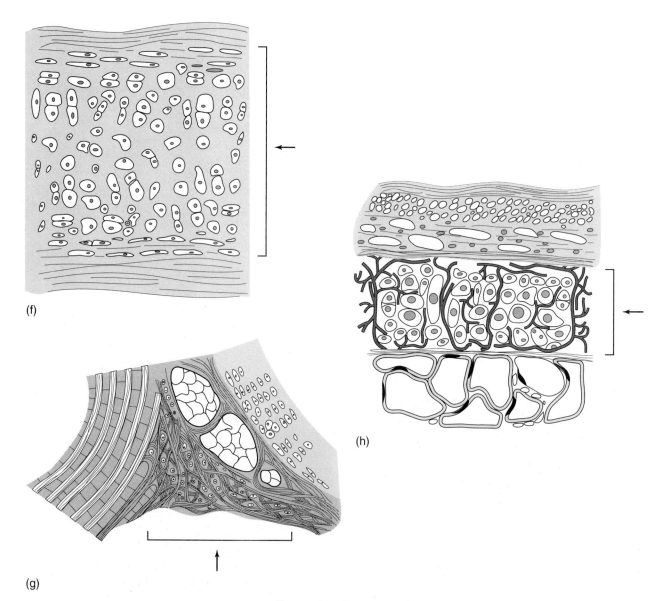

(f)

(g)

(h)

FIG. 3.4 Connective tissues, continued.

MICROSCOPE STUDY

Your instructor will provide prepared microscope slides of the epithelial and connective tissues. *Use* your textbook and *consult* Figures HA-2–17 to help you identify the components of the tissues. *Locate* the following. If you wish, *illustrate* the tissues on the next page or *add* to the labeling in Figures 3.1 through 3.4.

Epithelial Tissues
- simple squamous epithelium
    - nuclei
    - plasma membrane
    - cytoplasm
- simple cuboidal epithelium
    - nuclei
    - plasma membrane
    - cytoplasm
- simple columnar epithelium
    - nuclei
    - plasma membrane
    - cytoplasm
- stratified squamous epithelium
    - nuclei
    - plasma membrane
    - cytoplasm
- stratified cuboidal epithelium
    - nuclei
    - plasma membrane
    - cytoplasm
- stratified columnar epithelium
    - nuclei
    - plasma membrane
    - cytoplasm

Connective Tissues
- loose connective tissues
    - areolar connective
        - elastic fibers
        - collagen (KOL-lah-jen) fibers
        - fibroblasts (FĪ-brō-blasts)
- adipose
    - adipocyte (AD-ih-pō-sı̄t)
        - lipid (LIP-id) storage vacuole (VAK-u̠-ōl)
        - nucleus
        - plasma membrane
- reticular
    - reticulocyte (reh-TIK-u̠-lo̠-sı̠t)
    - reticular fiber
- dense connective tissues
    - dense regular
        - fibroblast
            - nucleus
        - collagen fibers
    - dense irregular
        - fibroblast
            - nucleus
        - collagen fibers
    - elastic connective
        - fibroblast
            - nucleus
        - elastic fiber
- cartilage
    - hyaline cartilage
        - lacunae (lah-KOO-nē)
        - chondrocyte (KON-dro̠-sı̠t)
            - nucleus
        - ground substance
    - fibrocartilage
        - lacunae
        - chondrocyte
            - nucleus
        - collagen fibers
    - elastic cartilage
        - lacunae
        - chondrocyte
            - nucleus
        - elastic fibers

LABORATORY REPORT

3

SHORT ANSWER

1. *Compare* and *contrast* epithelial and connective tissues.

2. A construction worker steps on a nail and drives it all the way to the bone. *List* the classes and types of tissue through which the nail passes from superficial to deep.

3. *Arrange* the following tissues from most cellular and least fibrous to least cellular and most fibrous:

   areolar connective, stratified squamous epithelium, adipose, dense regular

4. *List* each of the four tissue classes and provide a unique characteristic for each.

5. *List* the tissues that contain elastic fibers.

6. *List* the tissues that contain collagen fibers.

MATCHING

Match the structures in column A with the component tissue in column B.

A	B
_____ 1. tendon	a. elastic cartilage
_____ 2. epidermis	b. hyaline cartilage
_____ 3. lining of blood vessels	c. areolar
_____ 4. kidney tubule	d. reticular
_____ 5. salivary gland duct	e. stratified cuboidal epithelium
_____ 6. subcutaneous layer of skin	f. dense regular connective
_____ 7. yellow marrow	g. fibrocartilage
_____ 8. stroma of lymph nodes	h. adipose
_____ 9. tracheal cartilages	i. simple cuboidal epithelium
_____ 10. joint capsule	j. stratified squamous epithelium
_____ 11. epiglottis	k. dense irregular connective
_____ 12. intervertebral disk	l. simple squamous epithelium

Match the structures in column A with the component tissues in column B.

A	B
_____ 1. lines the intestine	a. areolar connective
_____ 2. contains chondrocytes and elastic fibers	b. elastic cartilage
_____ 3. contains elastic and collagen fibers	c. elastic connective
_____ 4. lines the vagina	d. hyaline cartilage
_____ 5. articular cartilages	e. nonkeratinized squamous epithelium
_____ 6. elastic arteries	f. simple columnar epithelium

MULTIPLE CHOICE

_____ 1. Which statement best describes connective tissue?
  a. Contains a large amount of matrix.
  b. Is primarily concerned with secretion.
  c. Lines an organ or body cavity.
  d. Contains tightly packed cells.

_____ 2. Areolar connective tissue is a type of
  a. dense connective tissue.
  b. cartilage.
  c. epithelial tissue.
  d. loose connective tissue.

_____ 3. Which of the following best defines an epithelial tissue?
  a. contractile   c. membrane
  b. conductive   d. supportive

_____ 4. In the scheme of organization of life, which of the following group to form tissues?
  a. cells   c. molecules
  b. organelles   d. organs

Integumentary System

Contents

Objectives

1. Define the integumentary system.
2. Name the layers and sublayers of the human skin and identify each under the microscope.
3. List a function for each layer and sublayer of the skin.
4. Name and identify each of the epidermal derivatives and provide a function for each.
5. Identify and name the parts of a fingernail.

DEFINITION OF INTEGUMENT

The integumentary (in-teg-u-MEN-tar-e) system is the first of the **organ systems** discussed in this manual. The **integument** consists of the epidermis, dermis, and sub-cutaneous layers. Each of these layers has specific functions. Some of the functions of the skin are thermoregulation; nourishment of the young; communication; exteroception; and protection from light, chemicals, physical damage, and infection from fungi and bacteria.

PERSPECTIVE VIEW OF HUMAN SKIN

Locate and *label* the following in Figure 4.1. For reference, see HA-19 and HA-20.

epidermis (ep-ih-DER-mis)

    stratum basale (STRA-tum ba-SAL)

    stratum spinosum (spi-NO-sum)

    stratum granulosum (gran-u-LO-sum)

    stratum lucidum (LU-sih-dum)

    stratum corneum (KOR-ne-um)

    epidermal derivatives (see HA18)

        sebaceous (se-BA-shus) gland

        sudoriferous (soo-do-RIF-er-us) gland

        hair

            follicle (FOL-lih-kl)

            bulb

arrector pili (ah-REK-tor pi-le) muscle (part of the dermis, not epidermis)

tarsal (TAR-sl) gland (see the eye, Chapter 16)

ceruminous (seh-ROO-mih-nus) gland (not illustrated)

dermis (DER-mis)

    papillary (PAP-ih-lar-e) layer

        dermal papillae

        blood vessels

        Meissner's corpuscles (MIS-ners KOR-pus-ls)

    reticular layer

        Pacinian (pah-SIN-e-an) corpuscles

subcutaneous (sub-kyoo-TA-ne-us) layer or hypodermis

    adipose tissue

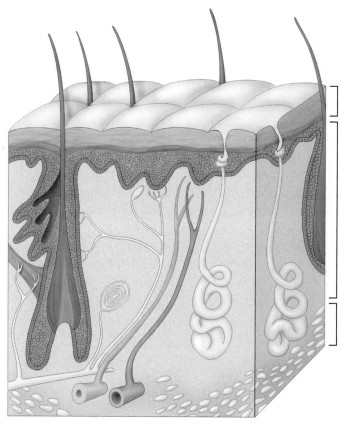

FIG. 4.1 Human skin, perspective view.

BREAST, RIGHT SAGITTAL VIEW

Locate and *label* the following in Figure 4.2:

mammary (MAM-er-e) gland

>   lobe

>>   lobule

>   mammary duct

>   lactiferous (lak-TIF-er-us) sinus

>   lactiferous duct

>   nipple

>   areola (ah-RE-o-lah)

>   suspensory ligament (of Cooper)

MICROSCOPE STUDY

Your instructor will provide a microscope slide of the skin. *Locate* the features that you labeled on Figure 4.1. *Look* at both the hairy and hairless skin if possible. If you have difficulties, *consult* with your laboratory instructor.

THE FINGERNAIL

Locate each feature on your own nail. *Use* your textbook to guide you.

eponychium (ep-o-NIK-e-um) or cuticle

lunula (LOO-nyoo-lah)

nail bed

nail body

nail fold

nail free edge

nail root

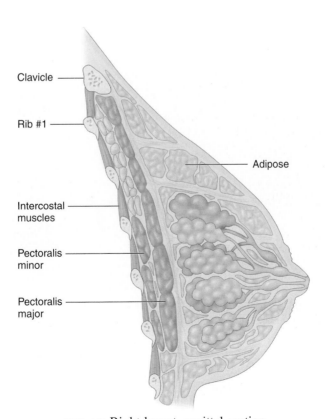

FIG. 4.2 Right breast, sagittal section.

SHORT ANSWER

1. In the space provided, *describe* several functions for each of the features listed. *Indicate* in which of the three primary skin layers each is located.

sebaceous gland

sudoriferous gland

hair

adipose tissue

stratum germinativum

stratum granulosum

stratum corneum

blood vessels

sensory receptors

papillary layer

stratum lucidum

reticular layer

2. *List* the layers and sublayers of the skin through which a needle would pass during a subcutaneous injection. *Begin* at the surface.

3. *Explain* why a suntan is lost over time.

4. *Describe* the structural and functional differences between an eccrine gland and a merocrine gland.

5. *Name* the gland that produces each of the following:

earwax

milk

sebum

pheromones

thermoregulatory sweat

6. *Evaluate* the following statement: "My grandmother was frightened so badly that her hair turned white overnight."

7. *Provide* a function for each of the following:

melanocytes

keratinocytes

Langerhans cells

Meissner's corpuscle

Pacinian corpuscle

stratum basale

8. *Evaluate* the following statement: "My aunt's skin is very wrinkled because she smokes and spends too much time in the sun."

MATCHING

Match the integumental feature in column B with the descriptions in column A.

A	**B**
_____ 1. forms friction ridges	a. mammary gland
_____ 2. mitotic layer	b. papillary layer, dermis
_____ 3. highly pigmented layer	c. reticular layer, dermis
_____ 4. contains elastic fibers and reticular fibers	d. stratum corneum
_____ 5. much adipose tissue	e. stratum germinativum
_____ 6. keratinized layer	f. stratum granulosum
_____ 7. products are NaCl, water, urea, and pheromones	g. sudoriferous gland
_____ 8. composed of alveolar glands	h. subcutaneous region

MULTIPLE CHOICE

_____ 1. Which of the following sequences is correct?

    a. epidermis, reticular layer, subcutaneous layer, papillary layer

    b. epidermis, papillary layer, reticular layer, subcutaneous layer

    c. epidermis, reticular layer, papillary layer, subcutaneous layer

    d. epidermis, subcutaneous layer, reticular layer, papillary layer

    e. none of the above

_____ 2. Which of the following statements about the functions of skin is NOT true?

    a. It synthesizes vitamin D.

    b. It helps regulate body temperature.

    c. It is sensitive to external stimuli.

    d. It is completely permeable.

_____ 3. The semicircular, white portion of the nail is the

    a. free edge.   c. body.

    b. eponychium.   d. lunula.

_____ 4. Which of the following statements about the skin is NOT true?

    a. It synthesizes vitamin D.

    b. It helps regulate body temperature.

    c. It is sensitive to many stimuli.

    d. It is modified to nourish the young.

    e. It is used to absorb water and minerals.

_____ 5. The hidden portion of the nail is the

    a. free edge.   d. root.

    b. eponychium.   e. lunula.

    c. body.

Skeletal System: Tissues and Organization

Contents

Objectives

1. Describe bone tissue and provide the function for each component.

2. Describe the two processes of bone formation.

3. Provide examples of bones formed by each process of osteogenesis.

4. Differentiate compact and cancellous bone.

5. Describe the structure of a long bone and provide the function for each component.

6. Name and describe the different kinds of fractures.

7. Define the terms structure, bone, and feature.

8. List and define the terms that describe the features of bony elements.

9. Define the two divisions of the skeleton.

FUNCTIONS OF THE SKELETON

The skeleton serves several important functions in the human organism:

- support
- protection
- blood cell formation
- mineral storage

Perhaps the most obvious function of the skeleton is support. All of the organs and organ systems are hung from the skeletal framework. The skeletal muscles that move the skeleton must be firmly attached. The skeleton reflects these attachments by having spines, tuberosities, trochanters, grooves, fossae, and other features on which the muscles originate and insert. The more heavily muscled an individual is, the larger or more coarse these features will be. It is often possible to tell a male from a female skeleton on the basis of skeletal features.

Protection is also an important function. The cranium, thoracic cavity, pelvis, and neural canal each surrounds critical organs that, if damaged, would certainly lead to impairment or death. The ribs, for example, surround the lungs and heart to provide a barrier and at the same time are separate and movable to allow and assist breathing.

Blood cell formation—hemopoiesis (he-mo-poy-E-sis)—is a third major function of the skeleton. The red marrow of bones produces red blood cells and some white blood cells as well as platelets. Blood-forming functions will be considered in detail in Chapter 12.

Another important function of the skeleton is mineral storage. Precise levels of sodium, potassium, and calcium must be maintained in the blood and tissues at all times to ensure proper functioning of the nervous system, muscles, and other organ systems. The skeleton is able to store minerals when they are abundant and to release them when they are in short supply. These functions maintain proper blood and tissue levels of minerals at all times.

HISTOLOGY OF BONE

Bone is a complex connective tissue. To understand the functions described above, it is important to understand the histology.

Bone formation is called ossification (os-ih-fih-KA-shun). There are two patterns of formation:

1. **Endochondral** (en-do-KON-dral) **ossification**—bone formed within a previously existing cartilage (for example, the ulna).

2. **Intramembranous ossification**—bone formed on or within a membrane (for example, the frontal bone).

There are two arrangements of bone that occur both in endochondral bone and membrane bone.

1. **Compact bone** contains few spaces and forms the outer layer of all bones and the shafts of long bones. Osteons (OS-te-onz) are the structural units that make up compact bone. An osteon consists of the osteocytes, lamellae, central canal, and canaliculi.

2. **Cancellous** (KAN-seh-lus) **bone** is spongy in appearance and consists of an irregular lattice of thin trabeculae (trah-BEK-yoo-la) made up of osteocytes and lamellae but without the regular structure of the osteon. Among the trabeculae is red marrow where blood cells are produced. Blood vessels reach the osteocytes directly in cancellous bone. In compact bone, blood vessels are confined to the central canals of the osteons and materials move from osteocyte to osteocyte at the points of contact.

As with all connective tissues, bone contains widely separated cells suspended in a **matrix**. **Osteoprogenitor** (os-te-o-pro-GEN-ih-tor) **cells** are derived from the mesenchyme. Such cells can divide and develop into **osteoblasts** (OS-te-o-blasts). Osteoblasts produce bone (matrix) but can no longer divide. Once the osteoblasts mature, they are called **osteocytes** (OS-te-o-sits). These cells can neither produce bone matrix nor divide. **Osteoclasts** (OS-te-o-klasts) are thought to arise from the monocyte, a white blood cell. Unlike the osteoblasts, osteoclasts reabsorb bone.

Osteoblasts are thus necessary for bone deposition and for mineral storage whereas osteoclasts destroy bone matrix, releasing minerals.

Review the chapter on the histology and function of bone in your textbook, then *continue* with the exercise in the next section.

ANATOMY OF A LONG BONE

Your instructor may provide long bones that have been cut so you can observe the inside. However, unless they are fresh tissue, some of the following will not be seen.

Locate the following terms and *label* them in Figure 5.1. For reference see HA-21.

compact bone

    proximal and distal epiphysis (eh-PIF-ih-sis)

    diaphysis (di-AF-ih-sis)

    metaphysis (meh-TAF-ih-sis [contains epiphyseal plate])

    articular (ar-TIK-u-lar) cartilage

    periosteum (per-e-OS-te-um)

    medullary (MED-u-lar-e) cavity

    endosteum (en-DOS-te-um)

    yellow marrow (MAR-o)

    nutrient foramen (fo-RA-men)

    perforating (Volkmann's [VOLK-mans]) canal

    osteon

        central (Haversian [hah-VER-shan]) canal

        lacunae (lah-KOO-ne)

        osteocyte

        lamellae (lah-MEL-e)

        canaliculi (ka-nah-LIK-u-li)

cancellous (KAN-seh-lus) or spongy bone

    trabeculae (tra-BEK-u-le)

    red marrow

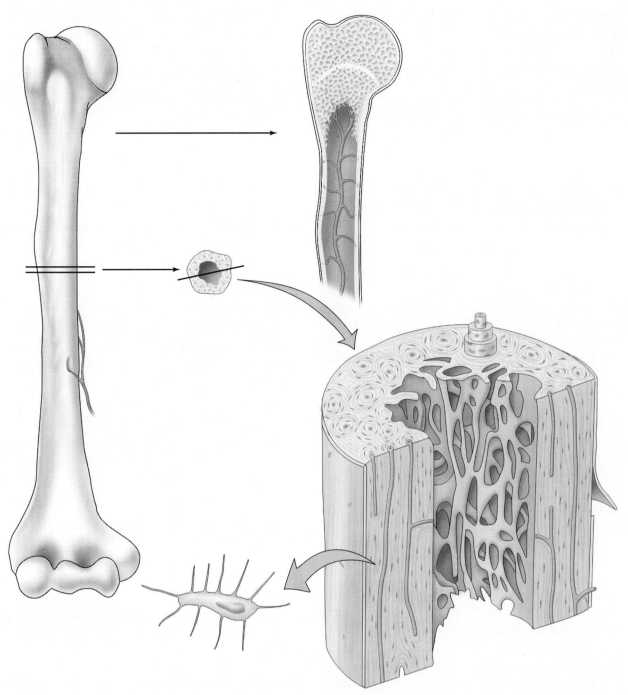

FIG. 5.1 Compact bone.

TERMINOLOGY OF SKELETAL ORGANIZATION

As you begin your study of the skeleton, *keep* the meanings of the following terms in mind.

> **Structure**—A unit made up of more than one bone such as the skull or os coxae.
>
> **Bone**—A skeletal element such as the humerus, temporal bone, or a metatarsal bone.
>
> **Feature**—A part of a bone such as the trochlea of the humerus, the medial malleolus of the tibia, or the perpendicular plate of the ethmoid.

During your study, you should *have* available a medical dictionary and a dictionary of word roots and combining forms. These references are found at your university or college bookstore or library. As an example, you might *use* them to look up the word *malleolus*, as used in the previous paragraph. In a dictionary of word roots you would find that the term *malle* is from the Latin meaning "a hammer." *Look* at the distal end of a tibia and observe that it is expanded and flattened, looking like the head of a small hammer. *Malle* is also the root of the word *mallet* as that used by a carpenter, judge, or club president.

As you continue your study of anatomy, *take* the time to determine the origin of the terms that you are attempting to learn. You will find that doing so is rewarding and beneficial to learning. Also, *be aware* that your textbook has an extensive glossary.

FEATURES OF BONY ELEMENTS

Memorize the definitions of the following terms before going on with the next section of this manual. The terms are used to describe the features of bony elements and are used frequently. If you are not sure of a meaning, *ask* your instructor to point out an example on the skeleton.

> **Canal** (ka-NAL)—A ductlike foramen such as the carotid canal.

> **Condyle** (KON-dil)—A knob or knucklelike projection that has a smooth surface for articulation with another bony element.
>
> **Crest**—A large ridge or elongated projection.
>
> **Facet** (FAS-et)—A smooth face. Usually where two bones articulate.
>
> **Fissure** (FISH-ur)—An elongated foramen such as the superior orbital fissure.
>
> **Fossa** (FOS-ah)—A trench or ditch that accepts an expanded organ such as the hypophyseal fossa that cradles the pituitary gland.
>
> **Meatus** (me-A-tus)—A tubelike opening such as the external acoustic meatus.
>
> **Process**—A projection from the surface of a bone used for muscle attachment.
>
> **Spine**—A narrow-edged or pointed process.
>
> **Sulcus** (SUL-kus)—A deep and elongated groove.
>
> **Trochanter** (tro-KAN-ter)—From the Greek, meaning "a runner." The large, leverlike projections on the femur.
>
> **Tubercle** (TOO-ber-kul)—From the Latin word *tuber*, meaning a swelling. A small rounded swelling.
>
> **Tuberosity** (too-beh-ROS-ih-te)—A larger, rough swollen area. These areas are for muscle attachment.

DIVISIONS OF THE SKELETON

Two divisions make up the skeleton. The **axial** (AK-se-al) **skeleton** consists of the skull, vertebrae, sacrum, coccyx, ribs, sternum, hyoid, and auditory ossicles. The **appendicular** (ap-en-DIK-u-lar) **skeleton** is composed of the superior and inferior appendages and their girdles.

Figure 5.2 illustrates the anterior view of the skeleton. Using colored pencils or another method, *mark* the two divisions of the skeleton.

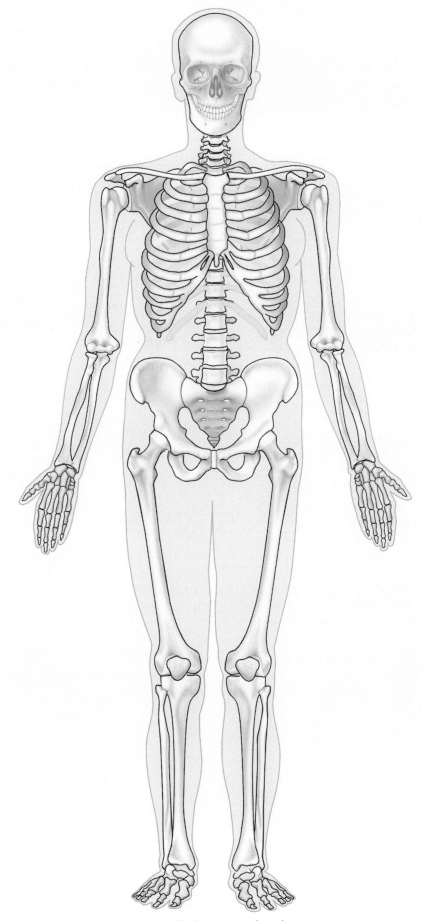

FIG. 5.2 Skeleton, anterior view.

LABORATORY REPORT

5

SHORT ANSWER

1. On the blank page opposite, *illustrate* the following fractures. *Describe* them below. You may *obtain* this information from your textbook.

partial

complete

displaced

closed (simple)

open (compound)

comminuted (KOM-ih-noo-ted)

greenstick

spiral

transverse

epiphyseal

impacted

2. *Describe* the following processes.
   intramembranous ossification

   endochondral ossification

   bone remodeling

3. Your instructor will supply a slide of ground bone. In the space below, *illustrate* and *label* your observations.

4. A stab wound has passed all the way to the medullary cavity of a long bone. *List* the layers through which the blade passed.

5. Young female gymnasts run a high risk of developing osteoporosis. Why?

MATCHING

Match the terms in column B with the descriptions in column A.

A	B
_____ 1. embryonic long bone	a. compact bone
_____ 2. osteolytic cell	b. chondrocyte
_____ 3. osteogenic cell	c. hyaline cartilage
_____ 4. cartilage-producing cell	d. fibrocyte
_____ 5. membrane around cartilage	e. perichondrium
_____ 6. fiber-producing cell	f. osteoblast
_____ 7. forms the epiphysis of long bones	g. osteoclast
_____ 8. forms the diaphysis of long bones	h. trabecular bone

Match the terms in column B with the descriptions in column A.

A	B
___ 1. smooth rounded articular process	a. canal
___ 2. elongated groove	b. condyle
___ 3. a round opening in a bone	c. fissure
___ 4. an elongated foramen	d. feature
___ 5. a broad trench or ditch	e. foramen
___ 6. a ductlike foramen	f. fossa
___ 7. a swollen area on the bone	g. spine
___ 8. a narrow pointed process	h. structure
___ 9. the general term for a part of a single bone	i. sulcus
___ 10. the general term for a unit consisting of more than one bone	j. tuberosity

MULTIPLE CHOICE

___ 1. Bone that forms first as cartilage is
   a. a flat bone.
   b. an endochondral bone.
   c. a membrane bone.
   d. the frontal bone.

___ 2. Which is a bone-producing cell?
   a. osteoblast   c. osteoclast
   b. osteocyte   d. chondrocyte

___ 3. A fracture in the shaft of a bone would be a break in the
   a. epiphysis.   c. articular cartilage.
   b. cancellous bone.   d. diaphysis.

___ 4. The portion of the long bone that stores yellow marrow in adults is the
   a. periosteum.   c. articular cartilage.
   b. endosteum.   d. medullary cavity.

___ 5. Indicate to which skeletal division the following belong. Use "A" for axial and "B" for appendicular.
   ___ clavicle

   ___ os coxae

   ___ manubrium of the sternum

   ___ frontal bone

   ___ hand

   ___ sacrum

   ___ ribs

   ___ hallux

___ 6. Which of the following is not a component of the skeleton?
   a. bone   c. skeletal muscle
   b. membranes   d. cartilage

___ 7. Bone grows in diameter at the
   a. medullary cavity.   c. perforating canal.
   b. Haversian canal.   d. periosteum.

___ 8. Which of the following is NOT a part of bony tissues?
   a. osteocyte.   c. central canal.
   b. canaliculi.   d. perichondrium.

___ 9. Bone grows in length at the
   a. medullary cavity.   c. pelvis.
   b. Haversian canal.   d. epiphysial disk.

___ 10. The two types of osseus tissues are
   a. cancellous and compact bone.
   b. Haversian and lamellar bone.
   c. trabecular and osteoclastic bone.
   d. spicular and trabecular bone.

___ 11. Two ways in which bone may form are
   a. stratified or simple.
   b. cuboidal or squamous.
   c. synovial or symphysis.
   d. endochondral or intramembranous.

_____ 12. The diameter of a developing bone enlarges as a result of

    a. the osteogenic periosteum.

    b. growth of the osteons.

    c. cartilage replacement.

    d. lamellar growth.

_____ 13. A fracture in which the bone portions remain in anatomical position is

    a. transverse.   c. nondisplaced.

    b. complete.   d. simple.

_____ 14. A comminuted fracture

    a. has many small pieces of bone.

    b. has only one side of the diaphysis affected.

    c. is of minor concern.

    d. is also called a Pott's fracture.

_____ 15. A Pott's fracture involves the

    a. tibial tuberosity.

    b. fibula.

    c. radius.

    d. None of the above are correct.

_____ 16. Processes that form where tendons attach to a bone include

    a. condyles and trochleas.

    b. fossae, sulci, and foramina.

    c. heads.

    d. trochanters, tuberosities, and tubercles.

_____ 17. A fracture produced by a twisting stresses is

    a. comminuted.   c. compression.

    b. Pott's.   d. spiral.

6 Skeletal System: Axial Skeleton

Contents

Objectives

1. Identify, name, and spell the bones and features of the cranium and face.

2. Identify, name, and spell the bones and features of the orbit and nasal cavity.

3. Identify, name, and spell the 32 permanent and 20 deciduous teeth.

4. Identify, name, and spell the features of the hyoid.

5. Identify, name, and spell the bones and features of the fetal skull.

6. Identify, name, and spell the vertebrae and features of the vertebral column.

7. Identify, name, and spell the bones and features of the thorax.

8. Differentiate a right rib from a left rib.

9. List and identify the structures of the axial skeleton.

HOW TO STUDY THE SKELETON

Label the diagrams on the pages that follow. *Use* your textbook as a reference. Each figure in this section is placed opposite a list of terms. *Write* the terms adjacent to the figure in your manual and *use* lines to lead to the items being labeled.

As in the previous chapters, the terms are presented in outline form. The outline indicates the relationship of

the items to be labeled. For example, look at the following list of terms:

cranial floor

    ethmoid bone

        crista galli

        cribriform plate

            olfactory foramina

The arrangment of this list indicates that the cranial floor is, in part, made up of the ethmoid bone. The ethmoid has two prominent features: the crista galli and the cribriform plate. The cribriform plate in turn contains the olfactory foramina.

Say each term aloud and then *write* it on a separate sheet of paper. *Use* your textbook, medical dictionary, or dictionary of word roots to *determine* the meaning of each term.

After labeling a figure, *refer* to the materials provided in your laboratory. *Use* your labeled figure as a study guide and *locate* each labeled item on the materials provided. When necessary, *refer* to your textbook for more information.

Make an effort to *locate* each item on the materials provided. If you are not able to do so in a reasonable amount of time, *consult with* your laboratory instructor or assistant. Don't forget to *use* your own body as a valuable learning resource. If a term is preceded by an asterisk (*), you can probably *see* or *feel* the feature on your own body. Making an effort to do this can help you remember the term.

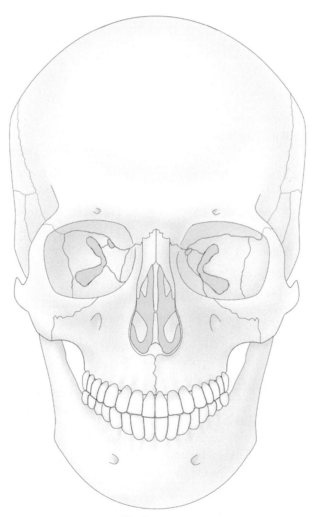

FIG. 6.1 Skull, anterior view.

Skull, Anterior View

Locate and *label* the following items in Figure 6.1.

*frontal (FRUN-tal) bone

*zygomatic (zi-go-MAT-ik) bone

    frontozygomatic suture (frun-to-zi-go-MAT-ik SOO-chur)

*maxillary (MAK-sih-ler-e) bone

    zygomaticomaxillary (zi-go-mat-ih-ko-MAK-sih-ler-e) suture

    intermaxillary suture

    *infraorbital (in-fra-OR-bih-tl) foramen

    *alveolar (al-VE-o-lar) process

*nasal (NA-zal) bone

    internasal suture

*mandible (MAN-dih-bl)

    *mental (MEN-tl) foramen

    *incisive (in-SI-siv) fossa

    *alveolar process

nasal cavity

    middle nasal conchae (KONG-ke)

    inferior nasal conchae

    nasal septum (SEP-tum)

        perpendicular plate of the ethmoid (ETH-moyd) bone

        vomer (VO-mer) bone

Orbit, Anterior View, Right

Locate and *label* the following items in Figure 6.2.

frontal bone (orbital process)

sphenoid (SFE-noyd) bone (orbital process)

    superior orbital fissure

    optic (OP-tik) foramen

zygomatic bone

maxillary bone (orbital process)

    inferior orbital fissure

ethmoid bone

*lacrimal (LAK-rih-mal) bone

    *lacrimal foramen

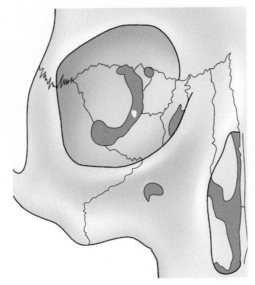

FIG. 6.2 Orbit, anterior view, right.

Skull, Right Lateral View

Locate and *label* the following items in Figure 6.3.

*parietal (pah-RI-eh-tal) bone

    sagittal suture (not illustrated)

    coronal (ko-RO-nl) suture

*occipital (ok-SIP-ih-tal) bone

    lambdoidal (lam-DOYD-al) suture

*sphenoid bone

    sphenoparietal suture

*temporal (TEM-po-ral) bone

    squamosal process (temporal squama)

    sphenosquamosal (sfe-no-skwa-MO-sl) suture

    squamosal suture

    *mastoid (MAS-toyd) process

    *external auditory (AW-dih-to-re) or acoustic (ah-KOOS-tik) meatus (me-A-tus)

    styloid (STI-loyd) process

    *zygomatic process

    temporozygomatic suture

*zygomatic bone

    *temporal process

*zygomatic arch (a structure consisting of the zygomatic and temporal processes meeting at the temporozygomatic suture)

    *mandible

    coronoid (KOR-o-noyd) process

    *condylar (KON-dih-lar) process

    *body

    *ramus (RA-mus)

    mandibular foramen

    *angle

Skull, Inferior View

Locate and *label* the following items in Figure 6.4.

maxillary bone

    dentition (den-TISH-un)

    *incisor (in-SI-zer) (There are four in the mandible, eight total.)

    *cuspid (KUS-pid) or canine (KA-nin) (There are two in the mandible, four total.)

    *premolar (pre-MO-lar) or bicuspid (There are four in the mandible, eight total.)

    *molar (There are six in the mandible, 12 total.)

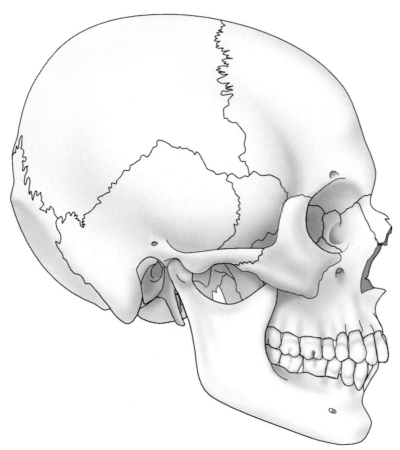

FIG. 6.3 Skull, right lateral view.

palatine (PAL-ah-tin) process

    anterior palatine foramen (incisive foramen)

*palatine bone

   median palatine suture

   transverse palatine suture

   greater palatine foramen

   lesser palatine foramen

vomer (VO-mer)

sphenoid

foramen ovale (O-val)

foramen spinosum (spi-NO-sum)

pterygoid (TER-ih-goyd) process

temporal bone

   *mandibular fossa (deflect your jaw laterally and palpate the opposite side)

   foramen lacerum (las-ER-um)

   carotid (kah-ROT-id) canal

   jugular (JUG-u-lar) foramen

   styloid process

   *mastoid process

occipital bone

   occipital condyle

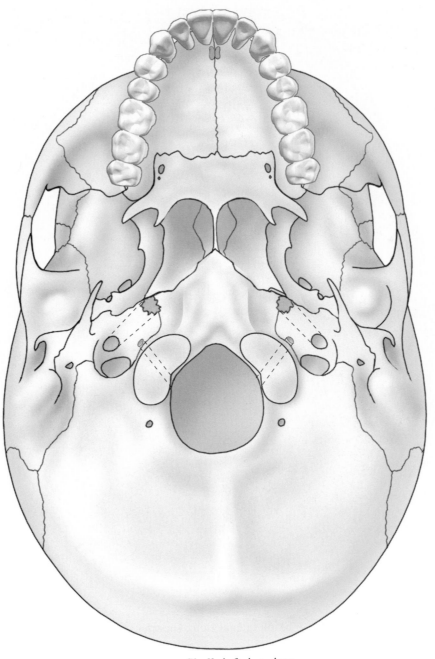

FIG. 6.4 Skull, inferior view.

condylar canal

hypoglossal (h<u>i</u>-p<u>o</u>-GLOS-al) canal

foramen magnum (MAG-num)

superior nuchal (NOO-kl) line

external occipital protuberance (pr<u>o</u>-TOO-ber-ans)

inferior nuchal line

frontal bone

ethmoid bone

   crista galli (KRIS-tah GAL-<u>e</u>)

   cribriform (KRIB-rih-f<u>or</u>m) plate

      olfactory (<u>ol</u>-FAK-t<u>o</u>-r<u>e</u>) foramina

sphenoid bone

   superior orbital fissure

   sella turcica (SEL-ah TUR-sih-kah)

      hypophyseal (h<u>i</u>-p<u>o</u>-FIZ-<u>e</u>-al) fossa

  foramen lacerum

Skull, Cranial Floor, Superior View

Locate and *label* the following items in Figure 6.5.

anterior, middle, and posterior cranial fossae

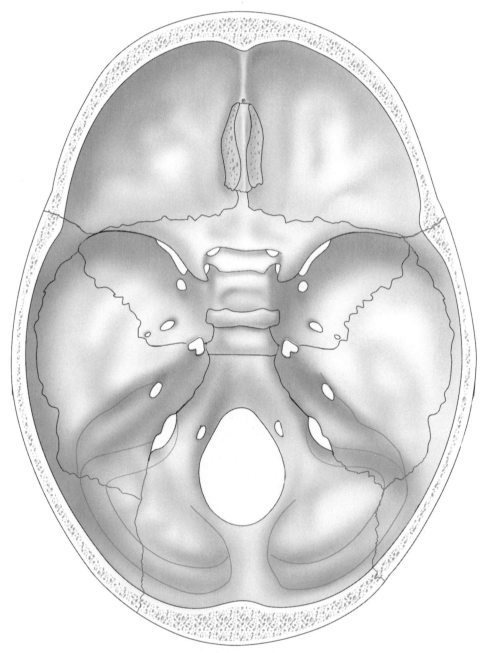

FIG. 6.5 Skull, cranial floor, superior view.

foramen spinosum

foramen rotundum (ro-TUN-dum)

optic foramen

temporal bone

internal acoustic meatus

petrous (PET-rus) portion

occipital bone

foramen magnum

hypoglossal canal

transverse sulcus

sigmoid (SIG-moyd) sulcus

Skull, Paranasal Sinuses, Lateral View

Locate and *label* the following items in Figure 6.6.

ethmoid sinus (SI-nus)

frontal sinus

maxillary sinus

sphenoid sinus

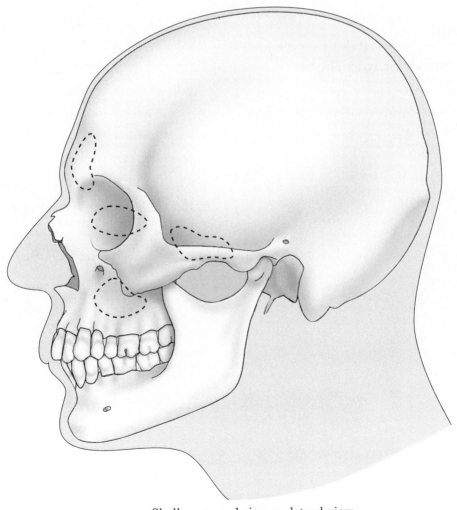

FIG. 6.6 Skull, paranasal sinuses, lateral view.

HYOID

Hyoid, Anterior View

Locate and *label* the following items in Figure 6.7.

*body of hyoid (H<u>I</u>-oyd)

  greater cornu (K<u>OR</u>-noo)

  lesser cornu

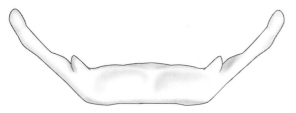

FIG. 6.7 Hyoid, anterior view.

FETAL SKULL

Fetal Skull, Superior and Right Lateral Views

Locate and *label* the following items in Figure 6.8.

fontanelle (fon-tah-NEL)

  anterior

  posterior

  posterolateral

  anterolateral

bones

  frontal

  parietal

  occipital

  temporal

  sphenoid

  maxillary

  mandible

sutures

  coronal

  sagittal

  lambdoidal

  squamosal

CRITICAL THINKING

For long answers, use a separate sheet of paper.

1. How is the adult dentition different from that of the juvenile?

2. Describe the bones and sutures involved in a cleft palate.

3. What two bones make up the nasal septum?

  _____

  _____

4. What six bones make up the orbit?

  _____

  _____

  _____

  _____

  _____

5. List the eight bones that make up the cranium.

  _____

  _____

  _____

  _____

  _____

  _____

  _____

  _____

6. List the 12 bones that make up the face.

  _____

  _____

  _____

  _____

  _____

  _____

  _____

  _____

  _____

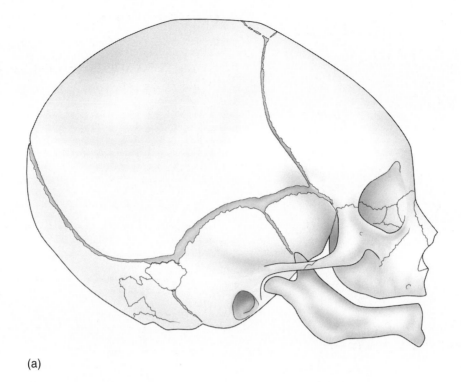

(a)

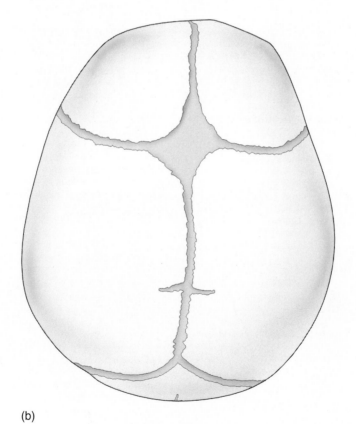

(b)

FIG. 6.8 Fetal skull, (a) right lateral and (b) superior views.

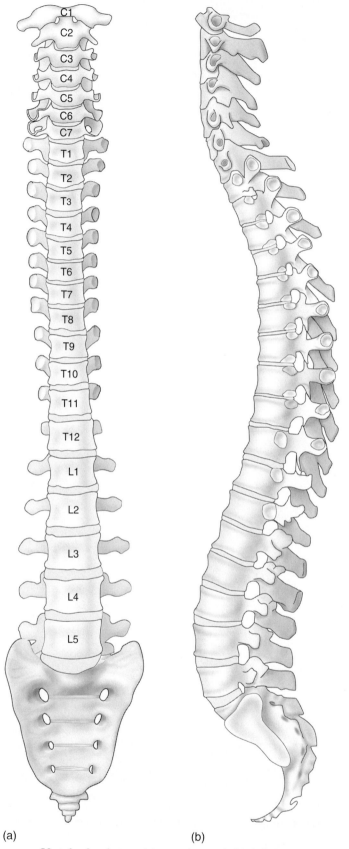

C1
C2
C3
C4
C5
C6
C7
T1
T2
T3
T4
T5
T6
T7
T8
T9
T10
T11
T12
L1
L2
L3
L4
L5

(a)   (b)

FIG. 6.9 Vertebral column, (a) anterior and (b) left lateral views.

7. List the major foramina of the skull and the bones in which they are found.

8. The medical examiner reports that a human mandible found by some hikers is that of a six- to ten-year-old child. How would the examiner know?

_____

_____

_____

_____

9. Which of the following does NOT contain a paranasal sinus?
   a. nasal
   b. ethmoid
   c. maxillary
   d. frontal

10. Which is NOT a feature of the temporal bone?
   a. external auditory meatus
   b. petrous portion
   c. mandibular fossa
   d. temporal process

VERTEBRAL COLUMN

Vertebral Column, Anterior and Left Lateral Views

Locate and *label* the following items in Figure 6.9.

*cervical vertebrae (SER-vih-kl VER-teh-bra) (there are 7)

*thoracic (tho-RAS-ik) vertebrae (there are 12)

*lumbar (LUM-bar) vertebrae (there are 5)

*sacral (SA-kral) vertebrae (there are 5)

*coccygeal (kok-SIJ-e-al) (usually 4)

intervertebral (in-ter-VER-teh-bral) disc

   nucleus pulposus (pul-PO-sus) (not illustrated)

   annulus fibrosus (AN-u-lus fi-BRO-sus) (not illustrated)

intervertebral foramen

cervical curve

thoracic curve

lumbar curve

sacral curve

Cervical Vertebrae, Superior and Right Lateral Views

Locate and *label* the following items in Figure 6.10.

body

vertebral foramen

vertebral arch

   pedicle (PED-ih-kl)

   lamina (LAM-ih-nah)

superior articular (ar-TIK-u-lar) process

   superior articular facet

inferior articular process

   inferior articular facet

transverse process

   transverse foramen

*spinous process

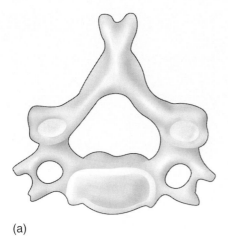

(a)

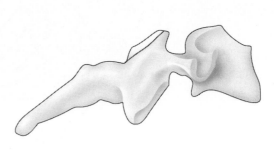

(b)

FIG. 6.10 Cervical vertebrae, (a) superior and (b) right lateral views.

Atlas (AT-las)—cervical vertebra 1 (C_1). Uniquely, the body of the atlas is lacking because it has been transferred to the axis. The large superior articular facets support the occipital condyles of the skull.

Axis (AKS-sis)—cervical vertebra 2 (C_2). An elongated body called the dens (denz) or odontoid (o-DON-toyd) process is the unique feature of this vertebra.

Atlas and Axis Perspective View

Locate and *label* the following items in Figure 6.11.

axis

    body

        dens (odontoid process)

    vertebral foramen

    vertebral arch

        pedicle

        lamina

    superior articular process

        superior articular facet

    inferior articular process

        inferior articular facet

    transverse process

        transverse foramen

    *spinous process

atlas

    vertebral foramen

    vertebral arch

        pedicle

        lamina

    superior articular process

        superior articular facet

    inferior articular process

        inferior articular facet

    transverse process

        transverse foramen

    *spinous process

Thoracic Vertebrae, Superior and Right Lateral Views

Locate and *label* the following items in Figure 6.12.

  body

  vertebral foramen

  vertebral arch

      pedicle

      lamina

  superior articular process

      superior articular facet

  inferior articular process

      inferior articular facet

  transverse process

      costal (KOS-tal) facet (T_1, T_{10}–T_{12}) or demifacet ([dem-e-FAS-et] T_1–T_9)

  *spinous process

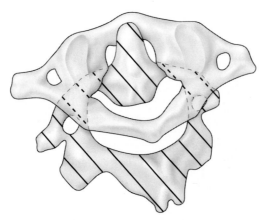

FIG. 6.11 Atlas and axis, perspective view.

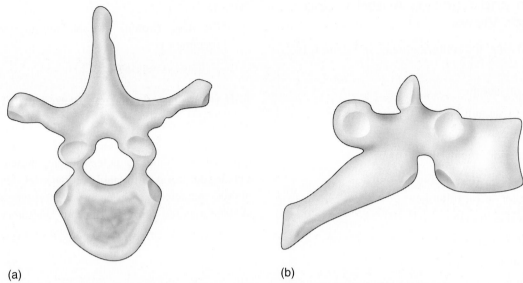

(a)   (b)

FIG. 6.12 Thoracic vertebrae, (a) superior and (b) right lateral views.

Lumbar Vertebrae, Superior and Right Lateral Views

Locate and *label* the following items in Figure 6.13.

body
vertebral foramen
vertebral arch
    pedicle
    lamina

superior articular process
    superior articular facet
inferior articular process
    inferior articular facet
transverse process
*spinous process

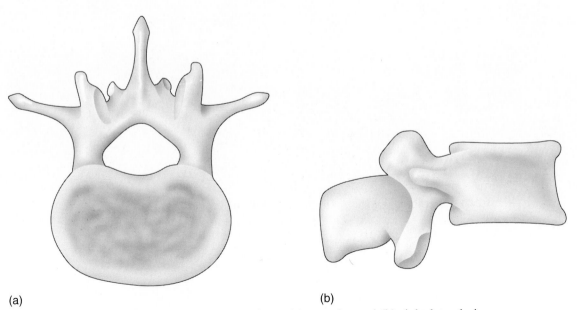

(a)   (b)

FIG. 6.13 Lumbar vertebrae, (a) superior and (b) right lateral views.

Sacrum and Coccyx, Anterior and Posterior Views

Locate and *label* the following items in Figure 6.14.

*sacrum

    base of sacrum

    apex of sacrum

    superior articular process

      superior articular facet

    anterior sacral (S<u>A</u>-kral) foramina

    posterior sacral foramina

    transverse lines

    sacral promontory

    median sacral crest

    auricular (aw-RIK-<u>u</u>-lar) surface

    sacral canal

*coccyx

    *Co_1–Co_4 (Some may be missing from your specimen.)

      transverse process of Co_1

Differences Among Vertebral Types

The typical characteristics vary within a type. However, there are characteristics that are unique to a type. These unique characteristics are shown in **UPPERCASE BOLD** in the table on page 69. *Locate* the unique characteristics of each vertebral type on the skeletal materials provided in your laboratory.

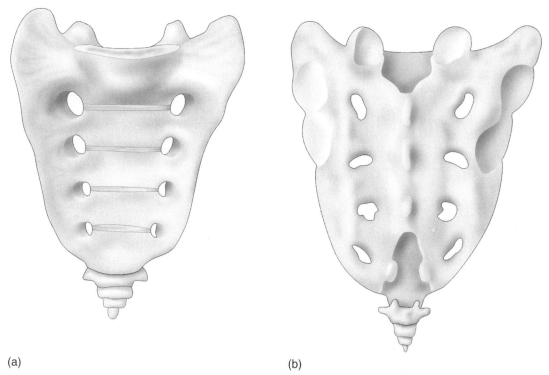

(a)   (b)

FIG. 6.14 Sacrum and coccyx, (a) anterior and (b) posterior views.

FEATURE	Lumbar	Thoracic	Cervical	Sacrum	Coccyx
Body	Largest	FACETS AND DEMIFACETS FOR RIBS	Smallest, raised lateral border	FIVE SACRAL VERTEBRAE SOLIDLY FUSED	USUALLY FOUR FUSED VERTEBRAE
Transverse process	Short and pointed	Long, FACET FOR RIB	short, TRANSVERSE FORAMEN	FUSED TO FORM SACRAL ALA	FIRST PAIR LARGE, THE REST ARE REDUCED
Spinous process	Large and square	Long, point inferior	Short, BIFID	FUSED TO FORM MEDIAN SACRAL CREST	ABSENT
Articular process	Large, facets in sagittal plane	Facets in frontal plane	Facets in transverse plane	One pair on the first sacral vertebrae	ABSENT

STERNUM

Sternum, Anterior View

Locate and *label* the following items in Figure 6.15(a).

*manubrium (mah-NOO-brē-um)

 *jugular or suprasternal notch

 clavicular (kla-VIK-u-lar) notch

 *sternal (STER-nl) angle

*body

 costal facet

 xiphoid (ZIF-oyd) process

RIBS

Ribs, Inferior View

Locate and *label* the following items in Figure 6.15(b).

head

tubercle (TOO-ber-kl)

body

 costal groove

Rib Types

Locate and *label* the following items in Figure 5.2.

ribs (1–12)

 *true ribs 1–7 (vertebrosternal)

 *false ribs 8–10 (vertebrochondral)

 *floating ribs 11, 12 (vertebral)

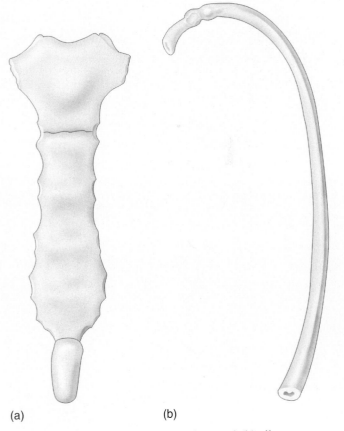

(a)   (b)

FIG. 6.15 Sternum, (a) anterior view and (b) rib, inferior view, right.

SHORT ANSWER

1. *List* the five vertebral types and the number of vertebrae in each group.

2. *List* the unique feature(s) of the following vertebral types:

   cervical

   thoracic

   sacral

3. *Name* the bones making up each of the following structures:

   cranium

   orbit

bony palate

nasal septum

sternum

MATCHING

Match the suture in column A with the bone it joins in column B.

A	B
_____ 1. lambdoidal	a. parietal bones
_____ 2. squamosal	b. nasal bones
_____ 3. coronal	c. temporal and zygomatic
_____ 4. sagittal	d. maxillary and palatine
_____ 5. sphenosquamosal	e. parietal and frontal
_____ 6. internasal	f. occipital and parietal
_____ 7. temperozygomatic	g. temporal and parietal
_____ 8. transverse palatine	h. sphenoid and temporal

MULTIPLE CHOICE

_____ 1. Which of the following belongs to the axial skeleton?

    a. manubrium   c. os coxae
    b. scapula   d. clavicle

_____ 2. Which is NOT part of the orbit?

    a. maxillary bone   c. vomer bone
    b. sphenoid bone   d. frontal bone

_____ 3. Which might NOT be part of the normal manus?

    a. navicular bone
    b. metacarpal bone
    c. pisiform bone
    d. distal phalanx of digit I of the manus

_____ 4. Which of the following forms an articulation with the axial skeleton?

    a. scapula   c. humerus
    b. ilium   d. none of the above

_____ 5. The term *petrous* means

    a. flat.   c. stone.
    b. temporal.   d. pituitary.

_____ 6. The term *hypophyseal* means

    a. flat.   c. stone.
    b. temporal.   d. pituitary.

_____ 7. Which of the following is characteristic of the male pelvis?

    a. The obturator foramen is triangular.

    b. The coccyx points into the pelvic outlet.

    c. The sacrum is broad and short.

    d. The ischial spines point posteriorly.

_____ 8. The bone of the cranium articulating with the atlas is the

    a. occipital.   c. axis.

    b. temporal.   d. sphenoid.

_____ 9. The palatine and a portion of what bone form the bony palate?

    a. mandible   c. occipital

    b. sphenoid   d. maxillary

_____ 10. The zygomatic arch is formed by processes of the

    a. frontal and maxillary.

    b. zygomatic and temporal.

    c. zygomatic and ethmoid.

    d. temporal and sphenoid.

_____ 11. In this lab manual, what does the asterisk (*) mean?

    a. Pay special attention to this feature.

    b. Locate this feature on your own body.

    c. You are not responsible for this feature.

    d. Observe this feature on laboratory materials.

_____ 12. Another name for the first cervical vertebrae is the

    a. coccyx.   c. axis.

    b. coxal.   d. atlas.

CHAPTER 7

Skeletal System: Appendicular Skeleton

Contents

Objectives

1. Identify, name, and spell the bones and features of the pectoral and pelvic girdles.

2. Identify, name, and spell the bones and features of the superior and inferior appendages.

3. Describe the differences in the male and female pelvis as related to the birth process.

4. List and identify the structures of the appendicular skeleton.

5. Differentiate right from left for all bones.

The appendicular division of the skeleton includes 126 bones. All of these bones are found associated with the limbs or their supporting girdles. Skeletal muscles are integrated with the bones to provide movements necessary for locomotion or for manipulation of objects.

As you learn the names of the bones, the structures they comprise, and their features, *determine* if the bone is left or right. Though for most bones this is not difficult, you may need your instructor's help with some.

Ask your instructor if you are to work with articulated hands and feet or if you must know the bones individually. *Remember* that a term preceded by an asterisk indicates that you are to *locate* the structure, bone, or feature on your own body.

Scapula (SKAP-yoo-lah), Posterior and Anterior Views

Locate and *label* the following in Figure 7.1(a) and (b).

*acromion (ah-KR<u>O</u>-m<u>e</u>-on) process

*coracoid (K<u>O</u>R-ah-k<u>o</u>yd) process

*superior border

*vertebral border

*axillary (AK-sih-l<u>ar</u>-<u>e</u>) border

  subscapular (sub-SKAP-<u>u</u>-lar) fossa

*scapular spine

*supraspinous (soo-prah-SP<u>I</u>-nus) fossa

*infraspinous fossa

  glenoid (GLEN-<u>o</u>yd) cavity

Clavicle (KLAV-ih-kl), Inferior View

Locate and *label* the following in Figure 7.2.

*sternal end

*acromial end

  coronoid tubercle

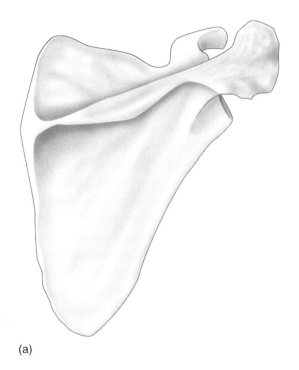

(a)

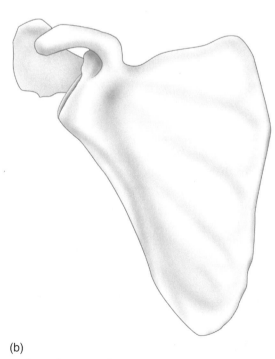

(b)

FIG. 7.1 Scapula, posterior (a) and anterior (b) views, right.

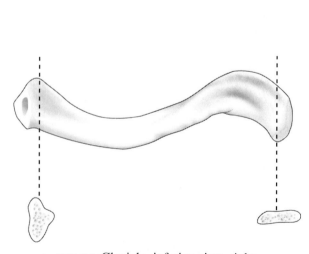

FIG. 7.2 Clavicle, inferior view, right.

SUPERIOR APPENDAGE

Humerus (HU-mer-us), Anterior and Posterior Views

Locate and *label* the following in Figure 7.3(a) and (b).

proximal head

greater tubercle

lesser tubercle

intertubercular groove

anatomical neck

surgical neck

*deltoid (DEL-toyd) tuberosity

*lateral and medial epicondyle (ep-ih-KON-dil)

capitulum (kah-PIT-u-lum)

trochlea (TROK-le-ah)

coronoid (KOR-o-noyd)

*olecranon (o-LEK-rah-non)

CRITICAL THINKING

1. What is the clinical significance of the surgical neck of the humerus?

    _____

    _____

    _____

    _____

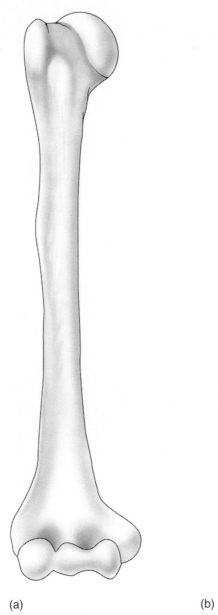

(a)

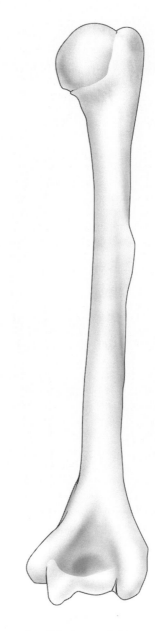

(b)

FIG. 7.3 Humerus, anterior (a) and posterior (b) views, right.

2. What is the function of the olecranon fossa?

_____

_____

_____

_____

3. Ask the students around you to hold an arm out in front of them, palm up, arm parallel to the floor, with the elbow fully extended. Do you notice a difference between males and females in the amount of extension at the elbow? If you do, explain the cause.

_____

_____

_____

Ulna (UL-nah), Anterior and Posterior Views

Locate and _label_ the following in Figure 7.4(a) and (b).

*olecranon process

 coronoid process

 trochlear notch

 radial (R<u>A</u>-d<u>e</u>-al) notch

 styloid (ST<u>I</u>-l<u>oy</u>d) process

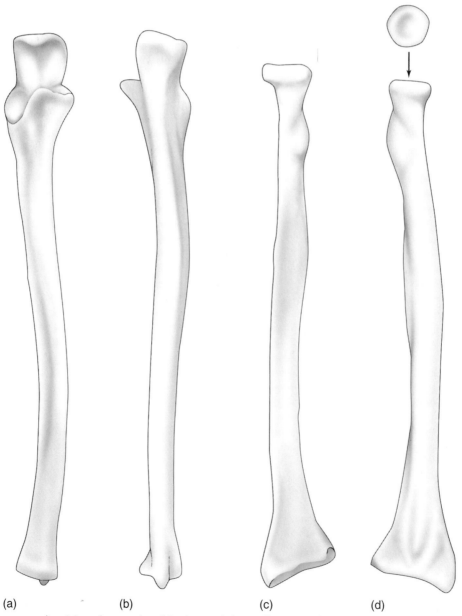

(a)   (b)   (c)   (d)

FIG. 7.4 Ulna, anterior (a) and posterior (b) views, right; radius, anterior (c) and posterior (d) views, right.

Radius (R<u>A</u>-d<u>e</u>-us), Anterior and Posterior Views

Locate and *label* the following in Figure 7.4(c) and (d).

head

radial tuberosity

ulnar notch

*styloid process

CRITICAL THINKING

Describe the location and appearance of a Colles' (K<u>OL</u>-ez) fracture.

_____

_____

_____

_____

Hand or Manus (MA-nus), Posterior View

Locate and *label* the following in Figure 7.5.

carpals (KAR-pls)

navicular (nah-VIK-<u>u</u>-lar) or scaphoid

lunate (LOO-n<u>at</u>)

triangular or triquetrum (tr<u>i</u>-KWEH-trum)

*pisiform (PIH-sih-f<u>o</u>rm)

greater multangular (mul-TANG-g<u>u</u>-lar) or trapezium

lesser multangular or trapezoid

capitate (KAP-ih-t<u>at</u>)

*hamate (HAM-<u>at</u>)

*metacarpals (met-ah-KAR-plz) (I–V)

*digits (DIJ-its) I–V

phalanges (fah-LAN-j<u>e</u>z)

*proximal row (I–V)

*middle row (II–V)

By convention the pollex is said to lack a middle phalanx, although anatomically it is the distal phalanx that is absent.

*distal row (I–V)

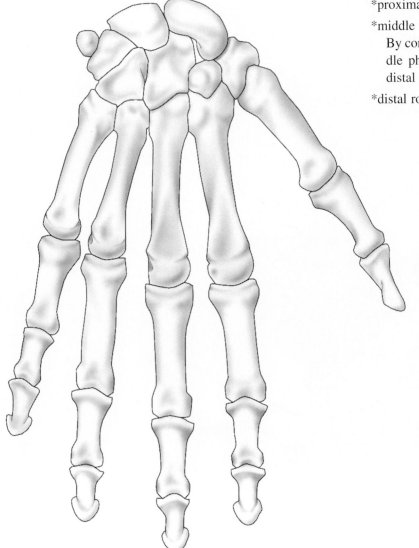

FIG. 7.5 Hand, posterior view, right

CRITICAL THINKING

1. Provide the common names for each digit.

_____

_____

_____

_____

2. What common word is derived from *manu*, the root of the word *manus?*

_____

_____

_____

_____

PELVIC GIRDLE

Bony Pelvis (PEL-vis), Anterior View

Locate and *label* the following in Figure 7.6.

coxal (KOK-sal) bones (os coxae)

sacrum [see again Figure 6.14(a) and (b)]

true pelvis

    pelvic inlet (superior aperture)

    pelvic outlet (inferior aperture)

pelvic brim

false pelvis

sacroiliac (sa-kro-IL-e-ak) joint

pubic symphysis (PYOO-bik SIM-fih-sis)

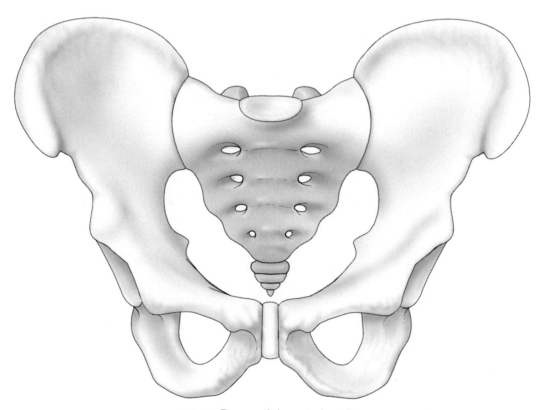

FIG. 7.6 Bony pelvis, anterior view.

Coxal Bones, Medial and Lateral Views

Locate and *label* the following in Figure 7.7(a) and (b).

ilium (IL-e-um)

    *iliac crest

    *anterior superior iliac spine

    anterior inferior iliac spine

    posterior superior iliac spine

    posterior inferior iliac spine

    *iliac fossa

    auricular surface

    greater sciatic (si-AT-ik) notch

ischium (IS-ke-um)

    ischial (IS-ke-al) spine

    *ischial tuberosity

pubis (PYOO-bis)

obturator (OB-tuh-ra-tor) foramen

acetabulum (as-eh-TAB-u-lum)

CRITICAL THINKING

1. Provide examples of the structural differences found in the male and female pelvis.

    _____

    _____

    _____

    _____

2. How do the various structural differences of the pelvis reflect the childbearing role of females?

    _____

    _____

    _____

    _____

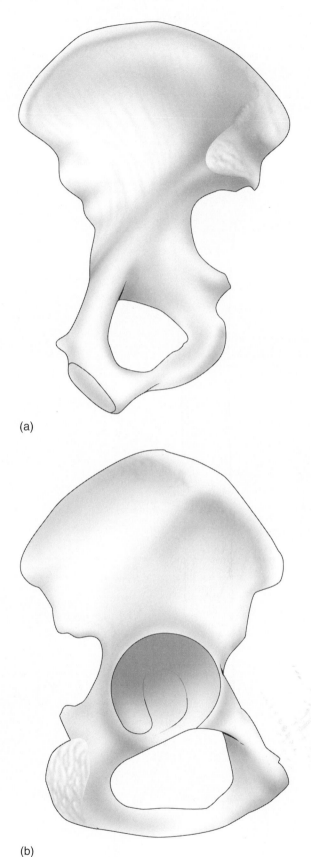

(a)

(b)

FIG. 7.7 Coxal bones, medial (a) and lateral (b) views, right.

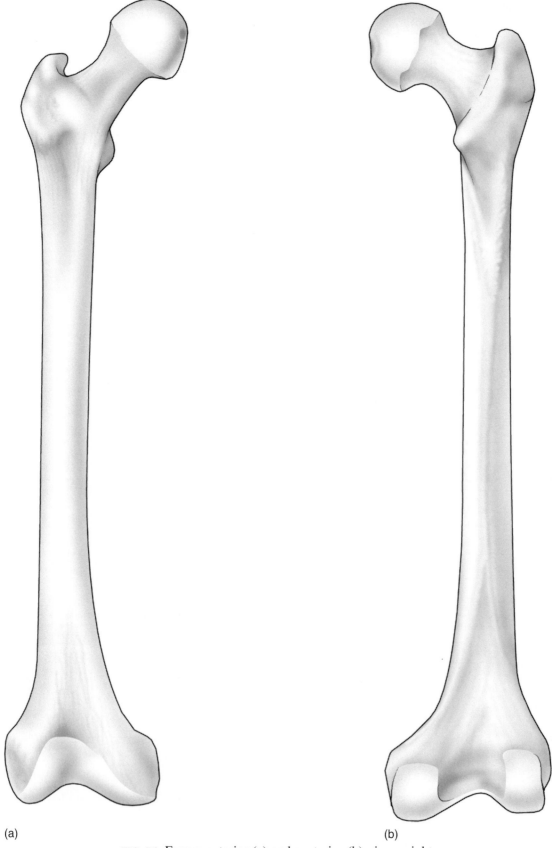

(a)   (b)

FIG. 7.8 Femur, anterior (a) and posterior (b) views, right.

INFERIOR APPENDAGE

Femur (FE-mur), Anterior and Posterior Views

Locate and *label* the following in Figure 7.8(a) and (b).

head

neck

*greater trochanter (tro-KAN-ter)

lesser trochanter

intertrochanteric crest

gluteal tuberosity (GLOO-te-al)

linea aspera (LIN-e-ah AS-per-ah)

*medial and lateral epicondyle (ep-ih-KON-dil)

medial and lateral condyle

intercondylar fossa

articular facet (FAS-et) for patella (pah-TEL-ah)

Tibia (TIB-e-ah), Anterior and Posterior Views

Locate and *label* the following in Figure 7.9(a) and (b).

lateral and medial condyle

intercondylar eminence

*tibial tuberosity

*medial malleolus (mal-LE-o-lus)

fibular (FIB-u-lar) notch

(a)   (b)

FIG. 7.9 Tibia, anterior (a) and posterior (b) views, right.

Fibula (FIB-u-lah), Lateral View

Locate and *label* the following in Figure 7.10(a).

*lateral malleolus

Patella, Anterior and Posterior Views

Locate and *label* the following in Figure 7.10(b) and (c).

*base

*apex

  articular facets

Foot (Pes), Superior and Medial Views

Locate and *label* the following in Figure 7.11.

  tarsal (TAR-sl) bones

    *talus (TA-lus)

    *calcaneus (kal-KA-ne-us)

    cuboid (KYOO-boyd)

    navicular (nah-VIK-u-lar)

    cuneiforms (KU-ne-ih-forms) I, II, III

*metatarsals (I-V)

  phalanges (fah-LAN-jez)

    *proximal row (I–V)

    *middle row (II–V) By convention the hallux is said to lack a middle phalanx, although anatomically it is the distal phalanx that is absent.

    *distal row (I–V)

*transverse and longitudinal arches

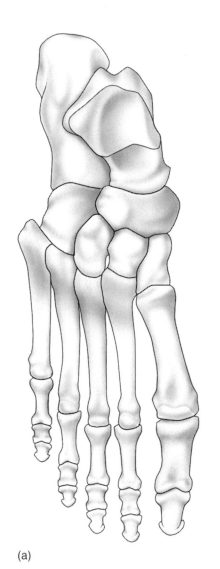

(a)

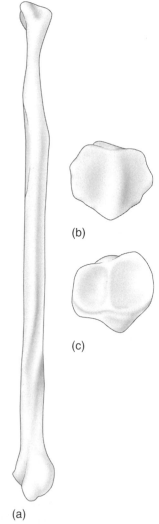

(b)

(c)

(a)

FIG. 7.10 Fibula, lateral view (a), right; patella, anterior (b) and posterior (c) views, right.

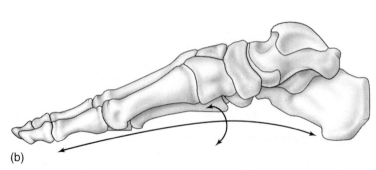

(b)

FIG. 7.11 Foot (pes), superior (a) and medial (b) views, right.

SHORT ANSWER

1. *Describe* a Pott's fracture.

2. *List* the skeletal components of a digit.

3. The most common fracture of the inferior appendage in the elderly is a broken "hip." *Describe* where and why this break occurs.

4. *Describe* how the patella is oriented relative to the femur and tibia.

5. *Name* the bones of the ankle.

MATCHING

Match the bone in column A with the feature in column B.

      A   **B**

_____ 1. ilium   a. acromion

_____ 2. femur   b. fibular notch

_____ 3. patella   c. intertubercular groove

_____ 4. scapula   d. apex

_____ 5. humerus   e. crest

_____ 6. tibia   f. linea aspera

_____ 7. ulna   g. radial olecranon process

MULTIPLE CHOICE

_____ 1. Which is NOT part of the manus?

     a. carpal
     b. metacarpal
     c. first middle phalange
     d. navicular

_____ 2. Which of the following structures consists of three bones?

     a. manus
     b. pes
     c. thorax
     d. os coxa

_____ 3. Which of the following applies to a left ulna when the bone is held in anatomical position?

     a. The styloid process is lateral.
     b. The medial malleolus is nearest to the mid-sagittal line.
     c. The radial notch faces left.
     d. The trochlear notch faces posterior.

CHAPTER

8 Joints

Contents

- Joint Defined
- Structural Classes of Joints
- Functional Classes of Joints

- Types of Synovial Joints
- Identifying Joints
- Synovial Joint Anatomy

8

CHAPTER

Objectives

1. Name and identify the three structural classes of joints and provide examples of each.

2. Name and identify the three functional classes of joints and provide examples of each.

3. Describe the structure of fibrous, cartilaginous, and synovial joints.

4. List and provide examples of each type of synovial joint.

5. Differentiate between intracapsular and extracapsular ligaments.

6. Describe the movements made by each type of synovial joint.

7. Differentiate between a dislocation and a sprain.

JOINT DEFINED

Joints (**articulatio** [ar-tik-u-LA-she-o]) are the places where two or more bones join. The study of joints is **arthrology** (ar-THROL-o-je) from the Greek *arthro* = joint. Joints may be categorized according to either their structure or the degree of movement.

STRUCTURAL CLASSES OF JOINTS

Fibrous—has no synovial cavity and is held together by fibrous connective tissue.

Cartilaginous—has no synovial cavity and is held together by cartilage.

87

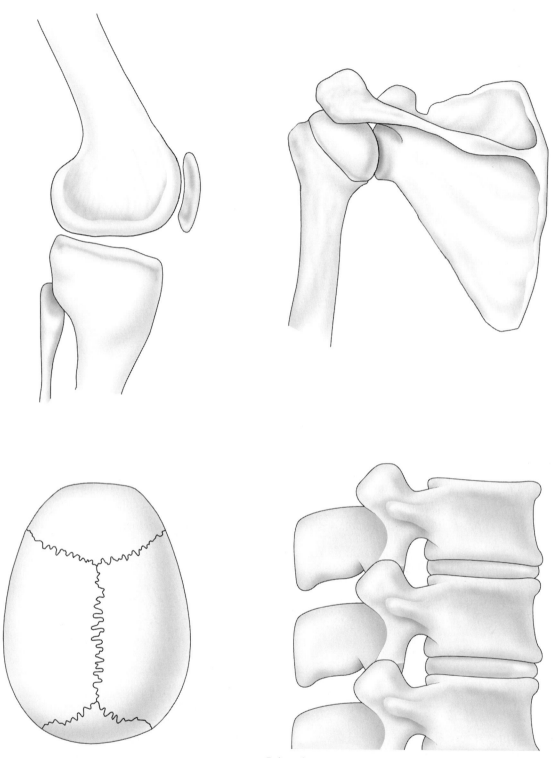

FIG. 8.1 Joint classes.

Synovial (sih-NO-ve-al)—synovial cavity is present and is held together by a fibrous capsule and accessory ligaments.

FUNCTIONAL CLASSES OF JOINTS

Synarthroses (sin-ar-THRO-sez) — immovable joints.

Amphiarthroses (am-fe-ar-THRO-sez)—slightly movable joints.

Diarthroses (di-ar-THRO-sez)—freely movable joints.

The synarthroses are represented by **sutures** (SOO-churs), fibrous joints found between the flat bones of the cranium, by **synchondroses** (sin-kon-DRO-sez), cartilaginous joints forming the epiphyseal plates of growing bones, and by the fibrous **gomphoses** (gom-FO-sez), of which the only example is the union between the roots of the teeth and the sockets formed by the periodontal ligament. Amphiarthroses include the **symphyses** (SIM-fi-sez), cartilaginous joints containing a pad of fibrocartilage such as the symphysis pubis and the intervertebral disks; and the **syndesmoses** (sin-des-MO-sez), fibrous joints where a band of ligament binds the two bones allowing partial movement, as in the tibiofibular joint. Diarthroses are all synovial. All synovial joints have the same basic components (below) but vary considerably in form.

TYPES OF SYNOVIAL JOINTS

The various types of synovial joints and examples of each are listed below. *Consult* your textbook for details on a particular joint.

Gliding—intercarpal and intertarsal joints.

Hinge—elbow joint.

Pivot—atlas to axis joint.

Condyloid—wrist.

Saddle—thumb (first metacarpal to carpal) joint.

Ball and socket—shoulder and hip joints.

IDENTIFYING JOINTS

Locate and *label* the following in Figure 8.1.

synarthrosis (suture)

amphiarthrosis (symphysis)

diarthrosis

    hinge

    ball-and-socket

SYNOVIAL JOINT ANATOMY

Locate and *label* the following in Figure 8.2.

articular cartilage

articular capsule

    fibrous capsule

    synovial membrane

synovial fluid

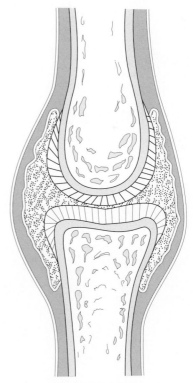

FIG. 8.2 Synovial joint anatomy.

SHORT ANSWER

1. For each of the following, *indicate* if it is a nonaxial, monaxial, biaxial, or triaxial joint. (You may need to *consult* your textbook.)

gliding

hinge

pivot

condyloid

saddle

ball-and-socket

2. Listed below are accessory ligaments of synovial joints. *Indicate* if they are extracapsular (E) or intracapsular (I).

_____ cruciate ligaments of the knee

_____ collateral ligaments of the knee

_____ sphenomandibular ligament of the mandible

_____ capitate ligament of the hip

_____ iliofemoral ligament of the hip

3. *Describe* the primary characteristics of each of the following:

fibrous joint

cartilaginous joint

synovial joints

4. *Describe* the following:

dislocation

sprain

5. *Describe* the anatomy and cause of nursemaid's elbow.

MULTIPLE CHOICE

_____ 1. In which joint are the bones not free to move relative to one another?

    a. symphysis

    b. synchondrosis

    c. condyloid

    d. amphiarthrosis

_____ 2. Which of these is NOT characteristic of a synovial joint?

    a. synovial membrane

    b. fibrous capsule

    c. articular cartilage

    d. immovable

_____ 3. A 45-year-old female has the following symptoms: inflammation of synovial membranes, accumulation of synovial fluid, pain and tenderness, and some immobility at certain joints. She probably has

    a. rheumatoid arthritis.

    b. intervertebral disc abnormality.

    c. gouty arthritis.

    d. bursitis.

_____ 4. A bursa

    a. acts as a joint.

    b. serves as a cushion.

    c. forms part of a bone.

    d. is a type of lever.

_____ 5. Which one of the following applies to a symphysis?

    a. synarthrosis

    b. short fibers between bones

    c. synovial

    d. fibrocartilaginous pad

_____ 6. Which of the following is multiaxial?

    a. hinge

    b. condyloid

    c. ball-and-socket

    d. pivot

_____ 7. The term applied to the forcible twisting of a joint with partial damage to its components _without_ luxation is

    a. strain.

    b. bursitis.

    c. dislocation.

    d. sprain.

_____ 8. A joint united by dense fibrous tissue that permits a slight degree of movement is a

    a. suture.

    b. gomphosis.

    c. synchondrosis.

    d. symphysis.

_____ 9. The deposition of sodium urate crystals in joints is

    a. osteoarthritis.

    b. gouty arthritis.

    c. rheumatoid arthritis.

    d. bursoarthritis.

_____ 10. Which of the following forms an articulation with the axial skeleton?

    a. scapula

    b. ilium

    c. humerus

    d. none of the above

_____ 11. Which of the following has an intracapsular ligament?

    a. hip joint

    b. knee joint

    c. elbow joint

    d. more than one of the above

9 Muscle Tissue

Contents

Objectives

1. Describe each type of muscle tissue and provide at least one location in the body.

2. List each level of muscle organization and describe its makeup. Include the name and location of the membrane surrounding each level.

3. Describe in detail the structure of the sarcomere.

4. Describe where and how muscle contraction occurs.

5. Define the terms origin, insertion, and action.

FUNCTIONS OF MUSCLE TISSUE

The muscles form the largest system of the human body. They make up the bulk of body weight and use most of the energy (food) that a person consumes. Thus the human is a "muscle animal"—that is an organism in which the other systems function to support the muscles. We eat to obtain energy to operate the muscles. A large portion of the brain is involved in sensing and controlling muscle movement and the vast majority of blood from each heartbeat is directed to the muscles.

All muscle tissues do work by contracting—that is, by shortening in length. Shortening is accomplished as the muscles convert chemical energy into mechanical energy. In addition to animating the skeleton, muscles maintain posture, move and regulate the movement of materials through hollow organs, and generate heat.

TYPES OF MUSCLE TISSUE

Three distinct types of muscle tissue are recognized: (1) **skeletal**, (2) **cardiac**, and (3) **visceral**. Skeletal muscles—the topic of this chapter—serve to animate the skeleton—thus their name. Cardiac muscle makes up the heart, and visceral muscle is associated with the digestive tract and the blood vessels. Cardiac and visceral muscle will be considered in later chapters.

HISTOLOGY OF MUSCLE TISSUE

Your instructor will provide microscope slides of each tissue type. *Use* your textbook and Figure 9.1 to *identify* the following:

skeletal muscle

    striations

    nucleus

visceral (VIS-er-al) **muscle**

    nucleus

cardiac (KAR-de-ak) **muscle**

    branching fibers

    intercalated disk

    nucleus

MUSCLE ULTRASTRUCTURE

The sarcomere (SAR-ko-mer) is the contractile unit of skeletal muscle. *Consult* Figure 9.2 and your textbook and then describe the structure and function of the following:

Z disk _____

_____

actin (AK-tin) _____

_____

myosin (MI-o-sin) _____

_____

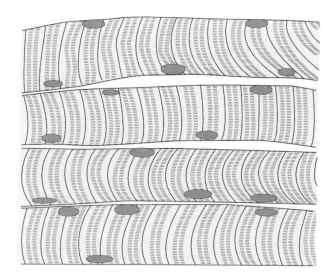

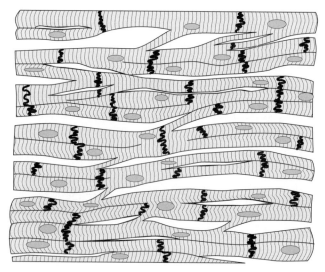

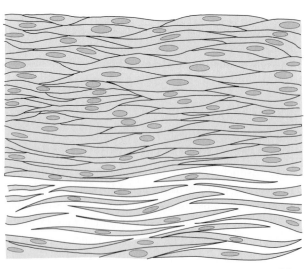

FIG. 9.1 Muscle tissue types.

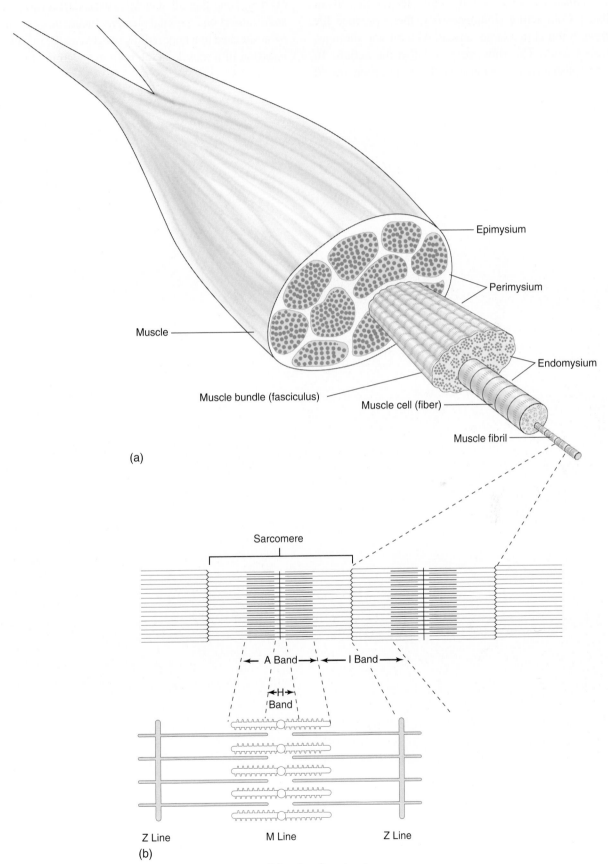

(a)

Epimysium

Perimysium

Endomysium

Muscle

Muscle bundle (fasciculus)

Muscle cell (fiber)

Muscle fibril

Sarcomere

A Band

I Band

H Band

Z Line

M Line

Z Line

(b)

FIG. 9.2 Muscle ultrastructure.

Sarcomeres are arranged into **myofibrils** (mi-o-FI-brils), which in turn combine to form muscle **fibers**. As you see, a muscle is comprised of numerous muscle cells or fibers. Contracting simultaneously, these produce the strength found in a large muscle. As a muscle shortens, it does work. This movement is called the **action**. To accomplish an action, a muscle pulls the insertion toward the origin. The **origin** is the fixed end of a muscle and the **insertion** is the movable end where the action is exerted. Note that all skeletal muscles arise (originate) and/or insert on a skeletal element, ligament, or aponeurosis attached to a bone. Midway between the origin and insertion of a muscle is the portion called the **belly**.

SHORT ANSWER

1. *Explain* the need for numerous mitochondria in skeletal muscle tissue.

2. *Describe* the primary differences in shape and appearance of each muscle tissue when viewed under the microscope.

MATCHING

Match the description in column A with the term regarding skeletal muscle in column B.

A	B
_____ 1. what a muscle does	a. action
_____ 2. large protein molecules	b. epimysium
_____ 3. contractile unit of a muscle	c. fasciculus
_____ 4. membrane surrounding a muscle	d. insertion
_____ 5. connects muscle to bone	e. myofibril
_____ 6. where a muscle exerts its action	f. myofilament
_____ 7. made of sarcomeres	g. neuron
_____ 8. a muscle pulls the insertion toward this point	h. origin
_____ 9. a bundle of muscle fibers	i. sarcomere
_____ 10. a motor unit = muscle fibers + _____	j. tendon

MULTIPLE CHOICE

_____ 1. A visceral muscle is found in the
  a. wall of the heart.
  b. wall of the small intestine.
  c. jaw muscle.
  d. arm muscle.

_____ 2. The largest of the structures listed is the
  a. fiber.   c. myofibril.
  b. sarcomere.   d. myofilament.

_____ 3. The connective tissue sheath around a fasciculus is the
  a. epimysium.   c. perimysium.
  b. endomysium.   d. fascia.

_____ 4. Which is NOT an action resulting from contraction of muscle fibers?
  a. movement of mucus along the trachea
  b. movement of urine through the urinary tract
  c. movement of blood through blood vessels
  d. movement of the body while walking

_____ 5. The term _motor unit_ is applied to
  a. the union of a muscle tendon with the periosteum of the bone.
  b. the triad in a skeletal muscle fiber.
  c. connective tissue coverings around a muscle.
  d. a motor neuron and the muscle fibers it controls.

_____ 6. Which generalization about skeletal muscles is NOT true?
  a. Muscles produce movements by pulling on bones.
  b. During contraction, the two articulating bones move equally.
  c. Actions act on the bone at the muscle insertion.
  d. The tendon attachment to the more stationary bone is the origin.

_____ 7. Which of the following does NOT increase the number of actions produced by a single muscle?
  a. multiple joints crossed
  b. motor units
  c. fixation
  d. pinnate structure

_____ 8. One of the most distinguishing histological features of cardiac muscle tissue is its
  a. nuclei.   c. cytoplasm.
  b. intercalated disks. d. striations.

_____ 9. During a muscular contraction
  a. the origin usually moves toward the insertion.
  b. the action occurs at the origin.
  c. the insertion moves toward the origin.
  d. the bone is being pushed by the muscle.

_____ 10. A cord of connective tissue that attaches a skeletal muscle to the periosteum of bone is a(n)
  a. aponeurosis.   c. sarcolemma.
  b. deep fascia.   d. tendon.

_____ 11. The muscles that contract when opening the hand after forming a fist are
  a. on the posterior brachium.
  b. all located in the palm of the hand.
  c. on the posterior antebrachium.
  d. on the anterior antebrachium.

_____ 12. According to the sliding-filament hypothesis
  a. thick myofilaments move toward each other.
  b. thin myofilaments move toward each other.
  c. Z disks move away from one another.
  d. potassium ions are necessary for contraction.

10 Skeletal Muscles

Contents

Objectives

1. List and define each major muscle action. Describe the plane in which each acts.

2. Using models, skeletons, your own body, or a cadaver, locate, name, and spell each of the muscles listed in this chapter.

3. For each of the muscles, provide the origin, insertion, action, synergist, and antagonist.

4. Know the meaning of each muscle name.

STUDYING MUSCLES

During your study of the skeletal muscles, you will learn the name (and correct spelling) of each muscle, the location of the muscle as well as its origin, insertion, and action. You will also learn about the synergists and an-tagonists of each muscle. A powerful study technique is to create lists of muscles by organizing them in a way that makes sense to you. For example, make a list of synergists and antagonists or create a list of muscles by region. The more ways you find to organize the information, the more likely you are to learn it.

NAMING MUSCLES

Muscles are named after one or more of the following characteristics:

shape

location

origin

insertion

action

size

An example of a muscle named after its shape is the trapezius. This muscle has a trapezoidal shape and is the most superficial muscle of the upper back. Other examples of muscles named according to shape are serratus, rhomboideus, and semimembranosus. *Use* your dictionary of word roots to *determine* what these terms mean.

Muscles named after a location are common and include pectoralis (the chest) and orbicularis oris (encircling the mouth). *Determine* the meanings of the following:

gluteus _____

_____

occipitalis _____

_____

femoris _____

_____

Origins and insertions of muscles are often contained in the name of the muscle. There are many examples. *Determine* the meanings of the following:

sternocleidomastoid _____

_____

thyrohyoid _____

_____

coracobrachialis _____

_____

iliocostalis _____

_____

Muscle actions such as adductor magnus, an adductor of the thigh, are often used in naming a muscle. Note that the term *magnus* refers to the largest of the thigh adductor muscles and, thus, describes the size. Flexor digitorum superficialis flexes the fingers, and it is the more superficial of two large complex muscles that do so.

Vastus lateralis and gluteus maximus are examples of muscles named for their size. Be critical in your analysis of muscle names. For example, in the case of flexor digiti minimi, the word *minimi* refers to the fifth digit of the hand (the little finger), not to the size of the muscle.

As you study the muscles, don't overlook clues to shape, location, origin, insertion, action, and size contained in the muscle name. These are free and useful bits of information that will help you learn the locations and functions of muscles.

TERMINOLOGY OF MUSCLE ACTION

Figure 10.1 illustrates the major muscle actions. Using the definitions supplied below, *label* each diagram with the appropriate action.

Flexion (FLEK-shun)—reduces the angle between two bony elements. This action is parallel to the sagittal plane. An example is bending the knee.

Extension (ek-STEN-shun)—increases the angle between two bony elements parallel to the sagittal plane. An example is straightening the knee.

Adduction (ad-DUK-shun)—movement toward the midline of a structure. This action is parallel to the frontal plane. An example is the movement of the superior appendages as they travel toward the body on the return phase of a butterfly stroke while swimming.

Abduction (ab-DUK-shun)—movement away from the midline parallel to the frontal plane. An example is spreading the fingers.

Rotation—movement of a body part around its long axis. An example is the action of the hand and forearm when using a screwdriver.

Elevation—lifting or raising a part of the skeleton. An example is the action of the shoulder during a shrug.

Depression—lowering a part of the body. An example is the mandible when opening the mouth.

Protraction—drawing a part of the body such as the mandible anteriorly.

Retraction—drawing a part of the body such as the mandible posteriorly.

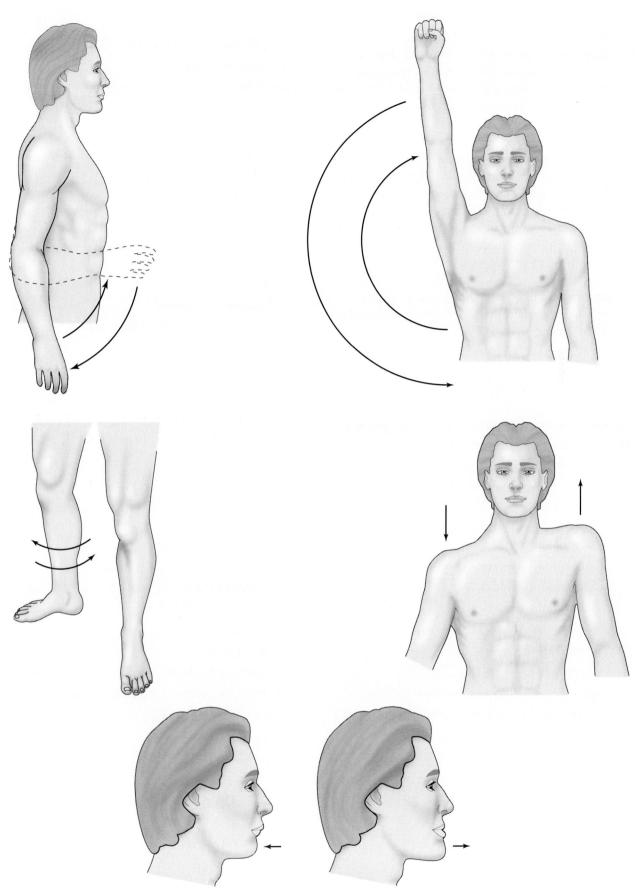

FIG. 10.1 Muscle actions.

ABOUT THE DIAGRAMS

Because muscles are complex and occur in layers at various depths, it is impossible to use just a single illustration to show all the muscles found in a given anatomical region. Notations are thus placed on the diagrams to indicate muscles that have been removed to promote clarity. The notation used to indicate such a removal is "ref."—an abbreviation for "reflected." Note the skeletal landmarks that have been labeled to help you with orientation. Read the figure captions. Is the view superior, inferior, medial, lateral, left, right, deep, superficial, etc.? Finally, read the accompanying description of each muscle on the facing page. Once you have learned the origin and insertion, you already know the action and location of that muscle.

If your instructor provides a cat for you to dissect, *turn* now to the introduction to Chapter 11 and *read* it before continuing.

MUSCLES OF FACIAL EXPRESSION

In the following lists of terminology, "o" refers to the origin, "i" to the insertion, and "a" to the action of the muscle.

Locate and *label* the following in Figure 10.2.

*frontalis (frun-TAL-is)
   o: galea aponeurotica (GAH-le-ah ap-o-nyoo-ROT-ik-ah)
   i: skin above the orbits
   a: elevates the eyebrow

*nasalis (na-ZAH-lis)
   o: maxilla
   i: aponeurosis of the nose and nasal cartilage
   a: compresses the external nares (nasal apertures)

*orbicularis oris (or-bik-u-LA-ris OR-is)
   o: fascia of muscles surrounding the mouth
   i: corner of the mouth
   a: compresses the lips

zygomaticus (zi-go-MAT-ih-kus) major
   o: zygomatic arch
   i: corner of the mouth
   a: elevates the corner of the mouth

*levator labii (LA-be-i) superioris
   o: maxilla below the orbit
   i: orbicularis oris (upper lip)
   a: elevates the superior lip

levator labii superioris alaeque nasi (AH-le-ke NA-zi)
   o: maxilla
   i: superior lip
   a: elevates the superior lip

levator anguli (ANG-gu-li) oris
   o: maxilla
   i: corner of the mouth
   a: elevates the corner of the mouth

*depressor anguli oris
   o: mandible
   i: corner of the mouth
   a: depresses the corner of the mouth

depressor labii inferioris
   o: mandible
   i: inferior lip
   a: depresses the inferior lip

mentalis (men-TAH-lis)
   o: mandible
   i: skin of the chin
   a: elevates the skin of the chin

*platysma (pla-TIZ-mah)
   o: fascia of deltoid and pectoralis major
   i: mandible and the corner of the mouth
   a: depresses the corner of the inferior lip and depresses the mandible

risorius (ri-ZOR-e-us)
   o: fascia of cheek
   i: corner of the mouth
   a: draws the corner of the mouth posteriorly

*buccinator (BUK-sih-na-tor)
   o: maxilla and mandible
   i: corner of the mouth
   a: compresses cheek

auricularis (aw-rik-u-LA-ris) anterior, posterior, and superior
   o: temporal fascia
   i: pinna (external ear)
   a: moves the pinna

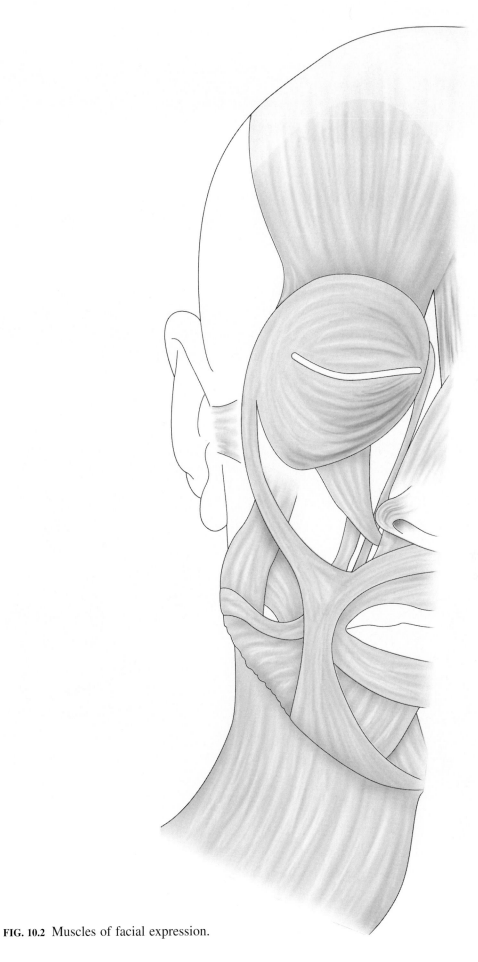

FIG. 10.2 Muscles of facial expression.

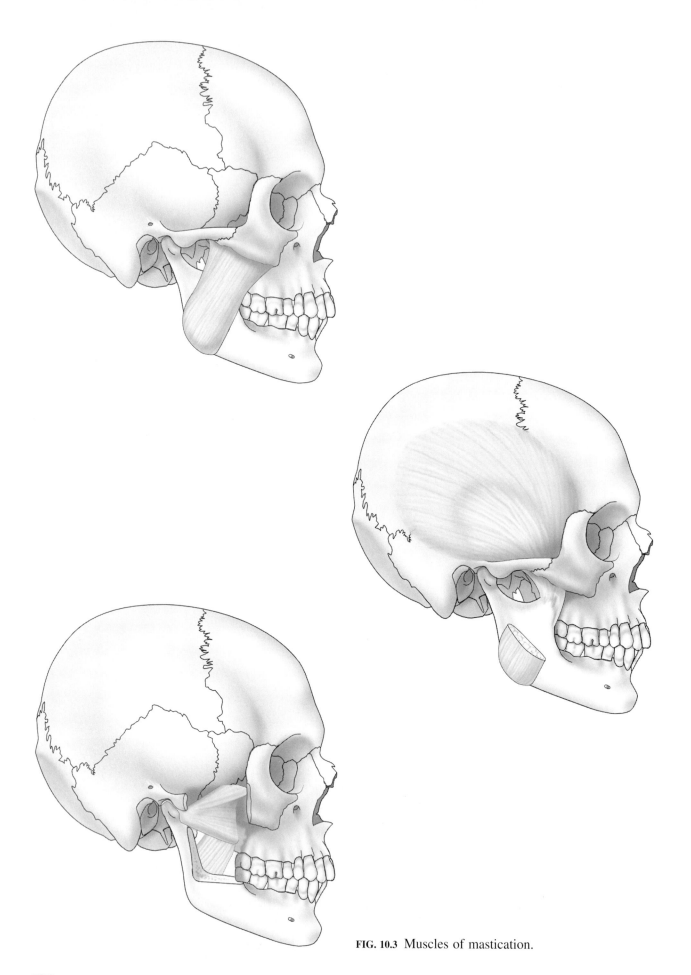

FIG. 10.3 Muscles of mastication.

*orbicularis oculi

    o: medial wall of the orbit

    i: skin around the eye

    a: closes the eye

procerus (pr<u>o</u>-S<u>E</u>-rus)

    o: bridge of the nose

    i: skin between the eyes

    a: depresses the eyebrows

*dilator (d<u>i</u>-L<u>A</u>-t<u>or</u>) nasi

    o: cartilage of the nose

    i: skin of the nose

    a: enlarges the external nares (nasal apertures)

MUSCLES OF MASTICATION (MAS-TIH-K<u>A</u>-SHUN)

Locate and *label* the following in Figure 10.3.

*masseter (mas-S<u>E</u>-ter)

    o: zygomatic arch

    i: angle of the mandible

    a: elevates the mandible

*temporalis ([tem-p<u>o</u>-RA-lis] deep to masseter)

    o: temporal bone

    i: coronoid process of the mandible

    a: elevates the mandible

medial pterygoid (TER-ih-g<u>o</u>yd)

    o: pterygoid process

    i: angle of the mandible

    a: elevates the mandible

lateral pterygoid (deep to the temporalis)

    o: pterygoid process

    i: condylar process of the mandible

    a: lateral movements of the mandible

MUSCLES OF THE HYOID

Locate and *label* the following in Figure 10.4.

digastricus (d<u>i</u>-GAS-trik-us)

    o: posterior belly on the mastoid process of the temporal bone; anterior belly on the mandible

    i: hyoid

    a: elevates the hyoid when both bellies act together: posterior belly retracts the hyoid; anterior belly protracts the hyoid

stylohyoideus (st<u>i</u>-l<u>o</u>-h<u>i</u>-<u>OY</u>-d<u>e</u>-us)

    o: styloid process of the temporal bone

    i: hyoid

    a: elevates and retracts the hyoid

mylohyoideus

    o: mandible

    i: hyoid

    a: elevates the hyoid

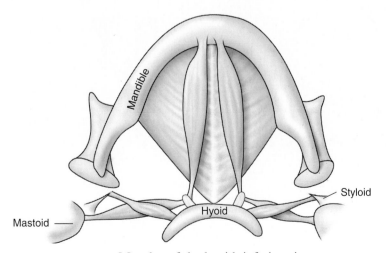

FIG. 10.4 Muscles of the hyoid, inferior view.

Locate and *label* the following in Figure 10.5.

geniohyoideus (je̱-ne̱-o-hi̱-OY-de̱-us)

    o: apex of the mandible

    i: hyoid

    a: protracts the hyoid

hyoglossus (hi̱-o̱-GLOS-us)

    o: hyoid

    i: lateral border of the tongue

    a: depresses the lateral border of the tongue

sternohyoideus (stern-no̱-hi̱-OY-de̱-us)

    o: clavicle and manubrium

    i: hyoid

    a: depresses the hyoid

omohyoideus (o̱-mo̱-hi̱-OY-de̱-us)

    o: superior border of the scapula and clavicle

    i: hyoid

    a: depresses the hyoid

thyrohyoideus (thi̱-ro̱-hi̱-OY-de̱-us)

    o: thyroid cartilage of the larynx

    i: hyoid

    a: depresses the hyoid and elevates the thyroid cartilage of the larynx

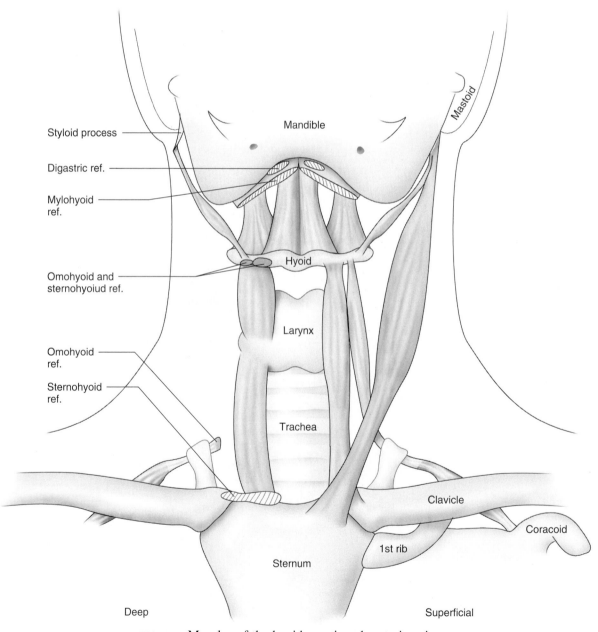

FIG. 10.5 Muscles of the hyoid, continued, anterior view.

MUSCLES OF THE ANTERIOR NECK

Locate and *label* the following in Figure 10.5.

sternothyroideus (stern-no-thi-ROY-de-us)

    o: manubrium

    i: thyroid cartilage of the larynx

    a: depresses the thyroid cartilage

sternocleidomastoideus (ster-no-kli-do-mas-TOY-de-us)

    o: clavicle and sternum

    i: mastoid process of the temporal bone

    a: elevates the face

MUSCLES OF THE PECTORAL GIRDLE

Muscles Acting on the Anterior Scapula

Locate and *label* the following in Figure 10.6.

*pectoralis (pek-to-RA-lis) minor (deep to pectoralis major)

    o: ribs 3–5

    i: coracoid process of the scapula

    a: depresses the scapula

serratus (ser-RA-tus) anterior (portions are deep to pectoralis minor)

    o: ribs 1–8

    i: vertebral border of the scapula

    a: abduction of the scapula and elevation of the ribs

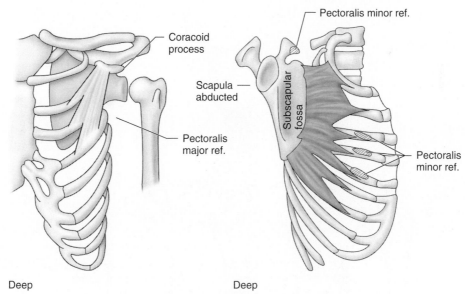

Deep   Deep

FIG. 10.6 Muscles acting on the anterior scapula.

Muscles Acting on the Posterior Scapula

Locate and *label* the following in Figure 10.7.

*trapezius (trah-PE-ze-us)

> o: occipital bone, nuchal ligament, spinous processes of C_7 and T_1–T_{12}
>
> i: scapular spine
>
> a: adducts, elevates, and depresses the scapula

levator scapulae (SKAP-u-la)

> o: transverse processes of C_1–C_4
>
> i: vertebral border of the scapula
>
> a: elevates the scapula

rhomboideus (rom-BOY-de-us) major

> o: spinous processes of C_7–T_1
>
> i: vertebral border of the scapula
>
> a: adducts the scapula

rhomboideus minor

> o: spinous processes of T_2–T_5
>
> i: vertebral border of the scapula
>
> a: adducts the scapula

MUSCLES OF THE SUPERIOR APPENDAGE

Muscles Acting on the Brachium, Anterior View

Locate and *label* the following in Figure 10.8.

*pectoralis major

> o: clavicle
>
> i: proximal humerus
>
> a: adducts the brachium

*deltoideus (del-TOY-de-us)

> o: clavicle and acromion process of the scapula
>
> i: deltoid tuberosity of the humerus
>
> a: abducts the brachium

coracobrachialis (kor-a-ko-bra-ke-AL-is)

> o: coracoid process of the scapula
>
> i: middle shaft of the humerus
>
> a: flexes and adducts the brachium

subscapularis (sub-skap-u-LAR-is)

> o: subscapular fossa
>
> i: lesser tubercle of the humerus
>
> a: medial rotation of the brachium

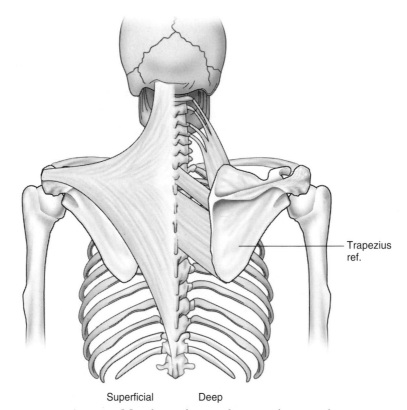

Trapezius
ref.

Superficial   Deep

FIG. 10.7 Muscles acting on the posterior scapula.

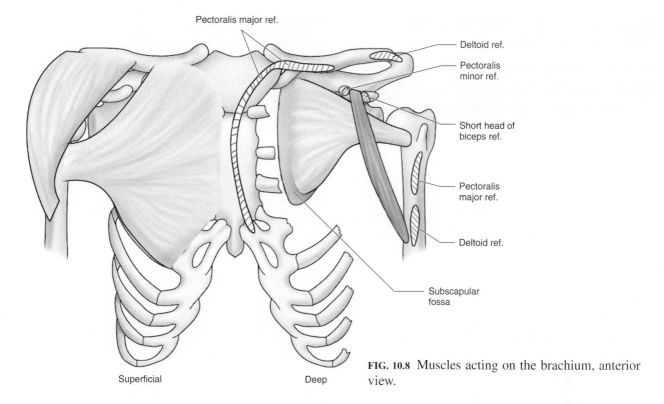

Pectoralis major ref.

Deltoid ref.

Pectoralis minor ref.

Short head of biceps ref.

Pectoralis major ref.

Deltoid ref.

Subscapular fossa

Superficial   Deep

FIG. 10.8 Muscles acting on the brachium, anterior view.

Muscles Acting on the Brachium, Posterior View

Locate and *label* the following in Figure 10.9.

*latissimus (la-TIS-i-mus) dorsi

  o: spinous processes of T_6–T_{12}, L_1–L_5, sacrum, and ilium

  i: intertubercular groove of humerus

  a: adducts and extends the brachium

*supraspinatus (soo-prah-spi-NA-tus)

  o: supraspinous fossa

  i: greater tubercle of the humerus

  a: abducts the brachium

*infraspinatus (in-fra-spi-NA-tus)

  o: infraspinous fossa

  i: greater tubercle of the humerus

  a: lateral rotation of the brachium

teres (TEH-rez) major (deep to the trapezius)

  o: inferior angle of the scapula

  i: distal to lesser tubercle of the humerus

  a: medial rotation of the brachium

teres minor (deep to the trapezius)

  o: axillary border of the scapula superior to the teres major

  i: greater tubercle of the humerus

  a: lateral rotation of the brachium

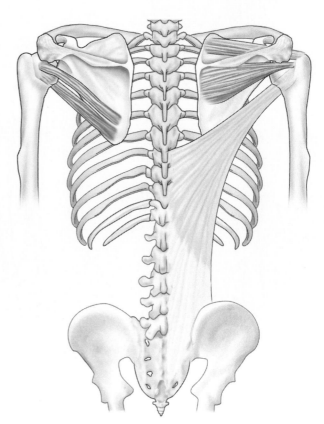

FIG. 10.9 Muscles acting on the brachium, posterior view.

MUSCLES OF THE SUPERIOR APPENDAGE

Muscles Acting on the Anterior Forearm (Flexors)

Locate and *label* the following in Figure 10.10.

*biceps brachii (BI-ceps BRA-ke-i)

    o: long head originates superior to the glenoid fossa and the short head originates on the coracoid process

    i: radial tuberosity

    a: flexes the forearm

brachialis (bra-ke-AL-is)

    o: anterior, middle humerus

    i: coronoid process of the ulna

    a: flexes the forearm

*brachioradialis (brak-e-o-ra-de-AH-lis)

    o: anterior, distal humerus

    i: distal radius

    a: flexes the forearm

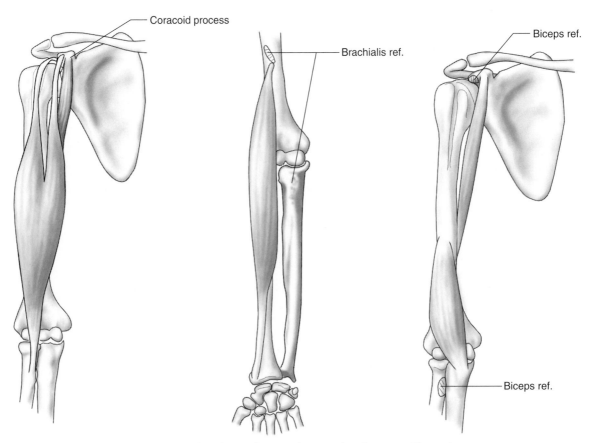

FIG. 10.10 Muscles acting on the anterior forearm (flexors).

Muscles Acting on the Posterior Forearm (Extensors)

Locate and *label* the following in Figure 10.11.

*triceps (TRI-ceps) brachii

   o: long head originates inferior to the glenoid fossa, the lateral head on the middle posterior humerus, and the short head on the proximal humerus

   i: olecranon process of the ulna

   a: extends the forearm

Locate and *label* the muscles acting on the antebrachium in Figure 10.12. *Use* the diagram to *learn* the relative positions of the muscles.

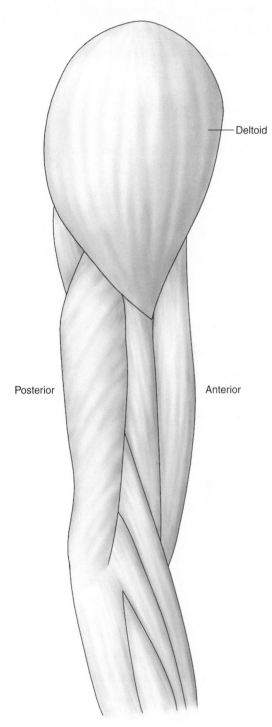

Deltoid

Posterior

Anterior

FIG. 10.12 Superficial muscles of the superior appendage, lateral view.

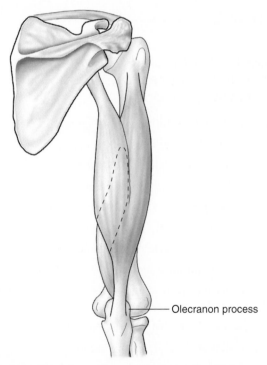

Olecranon process

FIG. 10.11 Muscles acting on the posterior forearm (extensors).

CRITICAL THINKING

1. Which is least related to the others regarding attachments of the muscle?

   a. insertion   c. movable end
   b. distal   d. fixed end

2. The name rhomboideus major tells you

   a. shape and location.
   b. shape and size.
   c. attachment and shape.
   d. location and size.

3. A muscle that opposes a particular movement is known as a

   a. fixator.   c. agonist.
   b. antagonist.   d. synergist.

4. Using a boat winch as an analogy, match the following:

   _____ boat

   _____ winch motor

   _____ attachment of the winch to the boat trailer

   _____ attachment of the winch to the boat

   a. origin   c. muscle
   b. insertion   d. skeletal element

5. List the muscles involved in biting down with great force.

   _____

   _____

6. List the muscles involved in a smile.

   _____

   _____

7. List the muscles involved in doing a "curl."

   _____

   _____

   _____

8. Provide the meanings of the following muscle names.

   biceps brachii _____

   _____

   levator labii superioris alaeque nasi _____

   _____

   sternocleidomastoideus _____

   _____

   digastricus _____

   _____

   palmaris longus _____

   _____

9. List the muscles acting on the hyoid as infrahyoidal and suprahyoidal.

   _____

   _____

   _____

   _____

   _____

   _____

   _____

   _____

Superficial Muscles Acting on the Anterior Forearm, Wrist, and Hand (Flexors)

Locate and *label* the following in Figure 10.13.

pronator (pro-NA-tor) teres

   o: medial epicondyle of the humerus

   i: midlateral surface of the radius

   a: medially rotates (pronates) the forearm

*flexor carpi radialis (KAR-pi ra-de-AH-lis)

   o: medial epicondyle of the humerus

   i: second and third metacarpal

   a: flexes the hand at the wrist, abducts the hand, and flexes the forearm

*palmaris longus (pal-MA-ris LONG-gus)

   o: medial epicondyle of the humerus

   i: palmar aponeurosis

   a: flexes the hand at the wrist and flexes the forearm

*flexor carpi ulnaris (ul-NA-ris)

   o: medial epicondyle of the humerus

   i: pisiform, hamate, and fifth metacarpal

   a: flexes the wrist

flexor digitorum superficialis (dig-ih-TOR-um soo-per-fish-e-AH-lis)

   o: medial epicondyle of humerus, the ulna, and the radius

   i: medial phalanges of digits 2–5

   a: flexes the medial phalanges of digits 2–5 and flexes the hand at the wrist

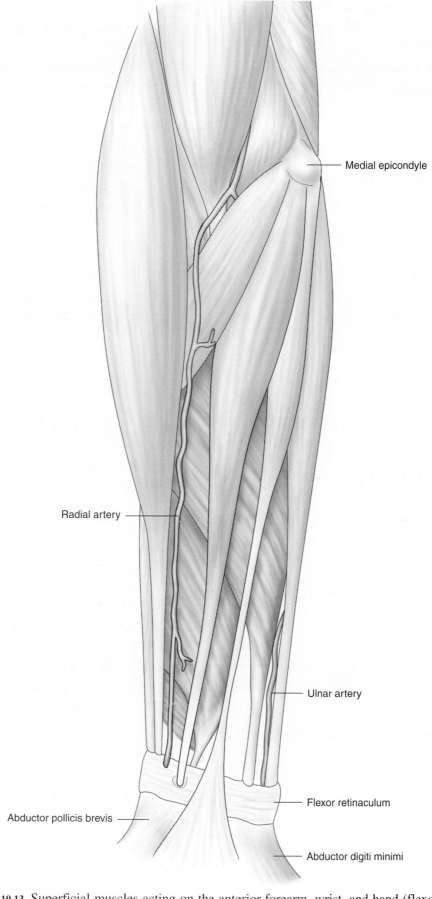

Medial epicondyle

Radial artery

Ulnar artery

Flexor retinaculum

Abductor pollicis brevis

Abductor digiti minimi

FIG. 10.13 Superficial muscles acting on the anterior forearm, wrist, and hand (flexors).

Deep Muscles Acting on the Anterior Forearm, Wrist, and Hand, Anterior (Flexors)

Locate and *label* the following in Figure 10.14.

supinator (SOO-pih-na̲-to̲r)

> o: lateral epicondyle of humerus
>
> i: lateral, proximal radius
>
> a: laterally rotates (supinates) the forearm

pronator quadratus (kwod-RA̲-tus)

> o: distal ulna
>
> i: distal radius
>
> a: medially rotates (pronates) the hand

flexor pollicis (PO̲L-is-cis) longus

> o: anterior radius and interosseus (in-ter-OS-e̲-us) membrane
>
> i: proximal distal phalanx of pollex (digit 1)
>
> a: flexes distal phalanx of pollex, abducts hand at wrist

flexor digitorum profundus

> o: ulna
>
> i: distal phalanges of digits 2–5
>
> a: flexes the distal and middle phalanges of digits 2–5 and flexes the hand at the wrist

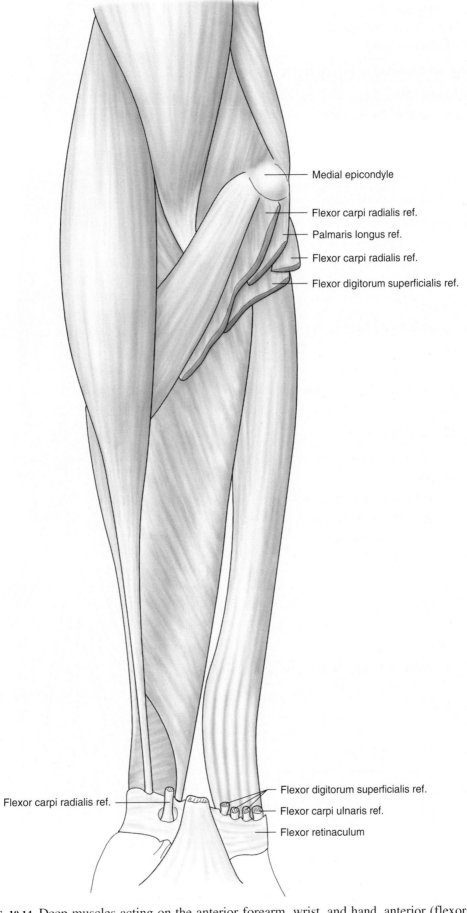

Medial epicondyle

Flexor carpi radialis ref.

Palmaris longus ref.

Flexor carpi radialis ref.

Flexor digitorum superficialis ref.

Flexor digitorum superficialis ref.

Flexor carpi ulnaris ref.

Flexor retinaculum

Flexor carpi radialis ref.

FIG. 10.14 Deep muscles acting on the anterior forearm, wrist, and hand, anterior (flexors).

MUSCLES OF THE SUPERIOR APPENDAGE

Muscles Acting on the Posterior Forearm, Wrist, and Hand (Extensors)

Locate and *label* the following in Figure 10.15.

anconeus (ang-KO-ne-us)

    o: lateral epicondyle of the humerus

    i: olecranon process of the ulna

    a: extends the forearm

extensor carpi radialis longus

    o: lateral epicondyle of the humerus

    i: second metacarpal

    a: extends and abducts the hand at the wrist

extensor carpi radialis brevis (BREV-is)

    o: lateral epicondyle of the humerus

    i: third metacarpal

    a: extends and abducts the hand at the wrist

extensor digitorum

    o: lateral epicondyle of the humerus

    i: middle and distal phalanges of digits 2–5

    a: extends digits 2–5 and extends hand at the wrist

extensor carpi ulnaris (ul-NA-ris)

    o: medial epicondyle of the humerus

    i: fifth metacarpal

    a: extends and adducts hand at the wrist

extensor digiti minimi (DIJ-ih-ti MIN-ih-mi)

    o: lateral epicondyle of the humerus

    i: proximal phalanx of digit 5

    a: extends digit 5

extensor pollicis longus

    o: lateral ulna

    i: distal phalanx of pollex

    a: extends the pollex and abducts the hand at the wrist

extensor pollicis brevis

    o: radius

    i: proximal phalanx of pollex

    a: extends the pollex and abducts hand at the wrist

abductor pollicis longus

    o: radius and ulna

    i: first metacarpal

    a: abducts the pollex and the hand at the wrist

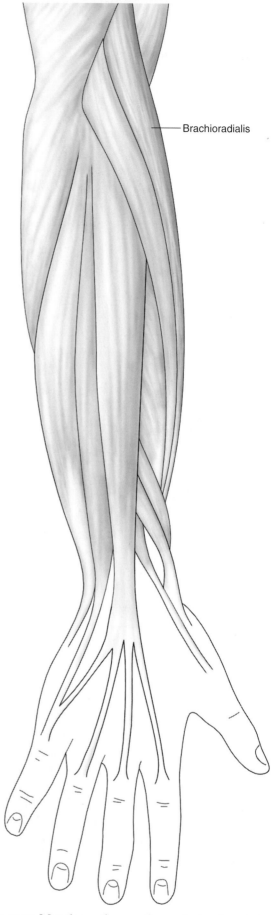

Brachioradialis

FIG. 10.15 Muscles acting on the posterior forearm, wrist, and hand (extensors).

Review Figure 10.16 to provide a general understanding of the location of the forearm muscles in cross section.

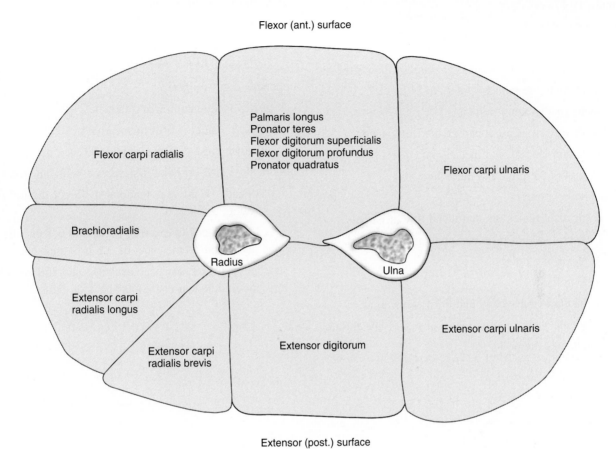

FIG. 10.16 Muscles of the forearm, cross section.

Intrinsic Muscles of the Hand, Anterior

Figure 10.17.

abductor pollicis brevis

    o: flexor retinaculum, lateral carpals

    i: lateral side of proximal phalanx of the pollex

    a: abducts the pollex

opponens (o-PO-nenz) pollicis

    o: flexor retinaculum and greater multangular

    i: lateral side of first metacarpal

    a: moves the pollex to meet digit 5

flexor pollicis brevis

    o: flexor retinaculum and distal row of carpals

    i: lateral side of proximal phalanx of pollex

    a: flexes the pollex

adductor pollicis

    o: capitate and second and third metacarpal

    i: medial side of proximal phalanx of the thumb

    a: adducts the pollex

abductor digiti minimi

    o: pisiform

    i: medial side of proximal phalanx of digit 5

    a: abducts and flexes digit 5

flexor digiti minimi brevis

    o: flexor retinaculum and hamate

    i: medial side of proximal phalanx of digit 5

    a: flexes digit 5

opponens digiti minimi

    o: flexor retinaculum and hamate

    i: medial side of fifth metacarpal

    a: moves digit 5 to meet the pollex

lumbricals (LUM-bri-kalz) first, second, third, and fourth

    o: outside of all four tendons of flexor digitorum (digits 2–5)

    i: to radial side of extensor digitorum of same digit just distal to the base of the proximal phalanx

    a: flexes the metacarpophalangeal joints and extends the middle and distal phalanges of digits 2–5

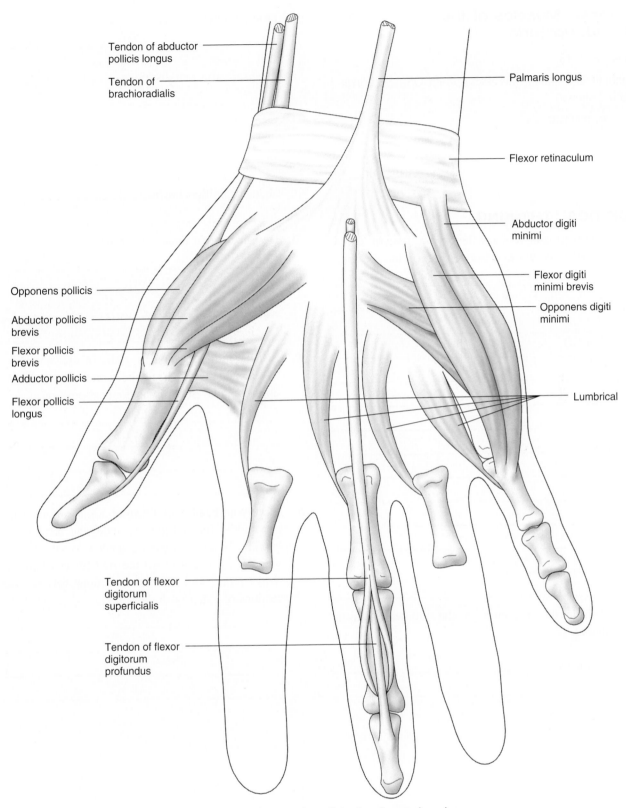

Tendon of abductor pollicis longus

Tendon of brachioradialis

Palmaris longus

Flexor retinaculum

Abductor digiti minimi

Opponens pollicis

Flexor digiti minimi brevis

Abductor pollicis brevis

Opponens digiti minimi

Flexor pollicis brevis

Adductor pollicis

Flexor pollicis longus

Lumbrical

Tendon of flexor digitorum superficialis

Tendon of flexor digitorum profundus

FIG. 10.17 Intrinsic muscles of the hand, anterior view.

Intrinsic Muscles of the Hand, Posterior

Figure 10.18.

palmar and dorsal interossei ([in-ter-OS-e-i], dorsal 1 and 2 shown)

  o: metacarpals

  i: proximal phalanges

  a: abduct digits 2–5

CRITICAL THINKING

1. List the major forearm flexors and forearm extensor (flexors and extensor located on the forearm).

   _____

   _____

   _____

   _____

   _____

   _____

   _____

   _____

   _____

   _____

   _____

   _____

2. Provide the meanings of the following muscle names.

   anconeus _____

   _____

pronator teres _____

_____

extensor carpi radialis longus _____

_____

interossei _____

_____

abductor pollicis brevis _____

_____

3. Describe carpal tunnel syndrome.

   _____

   _____

   _____

   _____

4. List the muscle groups, in order of use, necessary to reach behind the head and scratch the ear on the opposite side. Begin in anatomical position.

   _____

   _____

   _____

5. Place your hand palm down on a flat surface. Curl the middle finger (digit 3) under the hand while leaving the other fingers extended. What happens when you attempt to lift the ring finger (digit 4)? Why? (*Hint:* look at the connections between the tendons of extensor digitorum.)

   _____

   _____

   _____

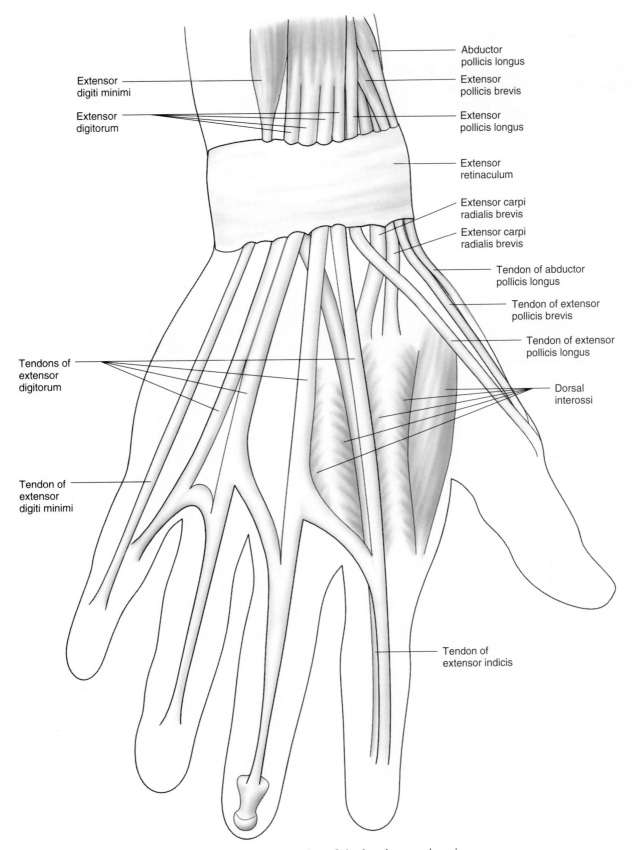

Extensor
digiti minimi

Extensor
digitorum

Abductor
pollicis longus

Extensor
pollicis brevis

Extensor
pollicis longus

Extensor
retinaculum

Extensor carpi
radialis brevis

Extensor carpi
radialis brevis

Tendon of abductor
pollicis longus

Tendon of extensor
pollicis brevis

Tendon of extensor
pollicis longus

Tendons of
extensor
digitorum

Dorsal
interossi

Tendon of
extensor
digiti minimi

Tendon of
extensor indicis

FIG. 10.18 Intrinsic muscles of the hand, posterior view.

MUSCLES OF THE SUPERIOR APPENDAGE

MUSCLES OF THE ANTERIOR TRUNK

Muscles of the Anterior Trunk, Superficial Two Layers

Locate and *label* the following in Figure 10.19.

*external/internal intercostalis (in-ter-kos-TAH-lis)

    o: ribs

    i: adjoining rib

    a: external elevates ribs; internal depresses ribs

*external oblique (o-BL<u>E</u>K)

    o: ribs

    i: linea alba

    a: compresses abdomen and lateral flexion of the spine

internal oblique

    o: ilium and vertebrae

    i: ribs

    a: compresses abdomen and lateral flexion of the spine

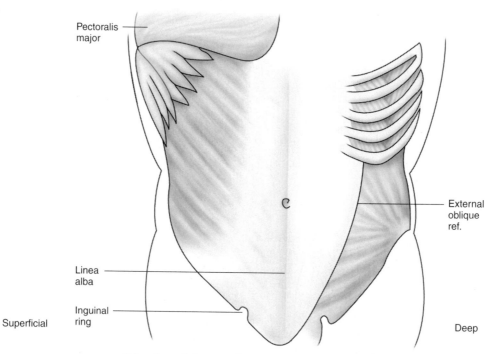

Pectoralis major

External oblique ref.

Linea alba

Inguinal ring

Superficial

Deep

FIG. 10.19 Muscles of the anterior trunk, superficial two layers.

Muscles of the Anterior Trunk, Deepest

Locate and *label* the following in Figure 10.20.

transversus abdominis (ab-DOM-ih-nis)

    o: ilium, ribs, and vertebrae

    i: sternum and linea alba

    a: compress abdomen

*rectus (REK-tus) abdominis

    o: pubis

    i: ribs and sternum

    a: flex vertebral column (as in a sit-up)

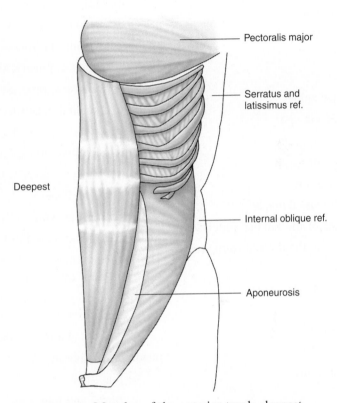

Deepest

Pectoralis major

Serratus and latissimus ref.

Internal oblique ref.

Aponeurosis

FIG. 10.20 Muscles of the anterior trunk, deepest.

MUSCLES OF THE POSTERIOR TRUNK

Deep Muscles Acting on the Vertebral Column and Skull

Locate and *label* the following in Figure 10.21.

*erector spinae group

iliocostalis lumborum (il-e-o-kos-TAH-lis lum-BO-rum)

  o: iliac crest

  i: inferior six ribs

  a: extends the lumbar spine

iliocostalis thoracis (thor-AS-sis)

  o: inferior six ribs

  i: superior six ribs

  a: stabilize the thoracic spine

iliocostalis cervicis (CER-vih-sis)

  o: superior six ribs

  i: transverse processes of C_4–C_6

  a: extends the cervical spine

longissimus (long-JIS-ih-mus) thoracis

  o: transverse processes of the lumbar vertebrae

  i: transverse processes of the thoracic vertebrae and ribs 9 and 10

  a: extends the thoracic spine

longissimus cervicis

  o: transverse processes of T_4–T_5

  i: transverse processes of C_2–C_6

  a: extends the cervical spine

longissimus capitis

  o: transverse processes of C_3–C_7 and T_1–T_5

  i: mastoid process

  a: extends and rotates the skull on the spine

spinalis (spi-NA-lis) thoracis

  o: spinous process of lumbar and inferior thoracic vertebrae

  i: spinous processes of superior thoracic vertebrae

  a: extends the thoracic spine

semispinalis thoracis (deep to spinalis)

  o: transverse process of T_6–T_{10}

  i: spinous processes of C_7–T_4

  a: extends and rotates the spine

semispinalis cervicis (deep to spinalis)

  o: transverse processes of T_1–T_5

  i: spinous process of C_2–C_5

  a: extends and rotates the cervical spine

semispinalis capitis (KAP-ih-tis)

  o: transverse processes of C_7 and T_1–T_6

  i: occipital bone (deep to splenius capitis)

  a: extends and rotates the head on the spine

splenius (SPLE-ne-us) capitis

  o: spinous processes of C_7 and T_1–T_3

  i: occipital bone

  a: rotates the skull on the spine

multifidus spinae (mul-TIF-ih-dus SPI-ne)

  o: sacrum and base of transverse processes of the cervical through the lumbar vertebrae

  i: spinous processes of the vertebrae that are two to four elements superior to the origin

  a: extends and rotates the spine

quadratus lumborum

  o: iliac crest

  i: rib 12 and transverse processes of L_1–L_4

  a: lateral flexion of the lumbar spine

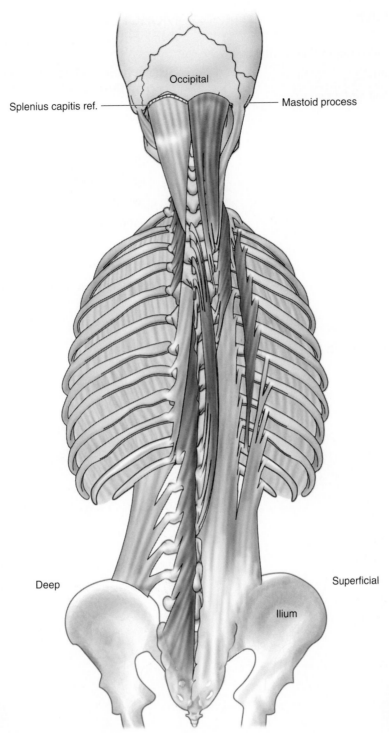

FIG. 10.21 Deep muscles of the posterior trunk acting on the vertebral column and skull.

THE DIAPHRAGM

Locate and *label* the following in Figure 10.22.

diaphragm (DĪ-ah-fram)

> o: ribs, sternum, and vertebrae
>
> i: central tendon
>
> a: depresses the central tendon (inhalation)

MUSCLES OF THE INFERIOR APPENDAGE

Superficial Muscles Acting on the Thigh, Posterior

Locate and *label* the following in Figure 10.23.

*gluteus (GLOO-tē-us) maximus

> o: posteror ilium and sacrum
>
> i: gluteal tuberosity
>
> a: extends and rotates the thigh

*gluteus medius/gluteus minimus

> o: ilium
>
> i: greater trochanter
>
> a: abducts and medially rotates the thigh at the hip

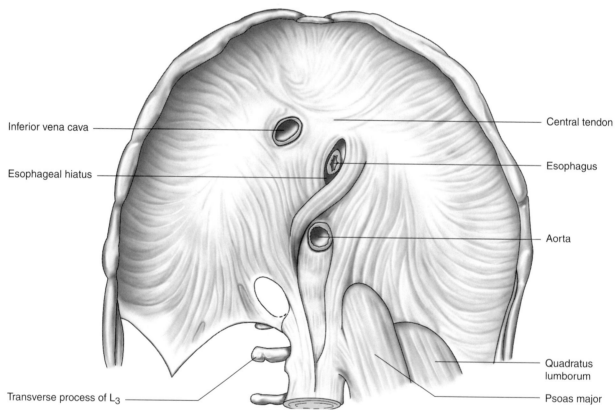

Inferior vena cava

Esophageal hiatus

Transverse process of L₃

Central tendon

Esophagus

Aorta

Quadratus lumborum

Psoas major

FIG. 10.22 Diaphragm, inferior view.

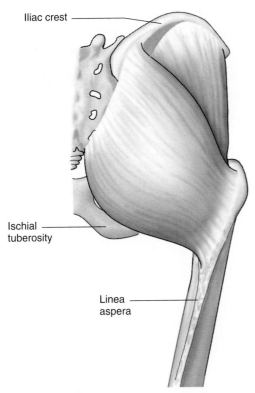

FIG. 10.23 Superficial muscles acting on the thigh, posterior view.

Superficial Muscles Acting on the Leg, Posterior

Locate and *label* the following in Figure 10.24.

*biceps femoris (B<u>I</u>-ceps FEM-<u>o</u>-ris)

   o: long head from the ischium
      short head from the linea aspera
      (LIN-<u>e</u>-a AS-per-a)

   i: proximal head of the fibula

   a: flexes the leg and extends the thigh

*semitendinosus (sem-<u>e</u>-ten-dih-N<u>O</u>-sus)

   o: ischium

   i: medial, proximal tibia

   a: flexes the leg and extends the thigh

*semimembranosus (sem-<u>e</u>-mem-brah-N<u>O</u>-sus)

   o: ischium

   i: medial, proximal tibia

   a: flexes the leg and extends the thigh

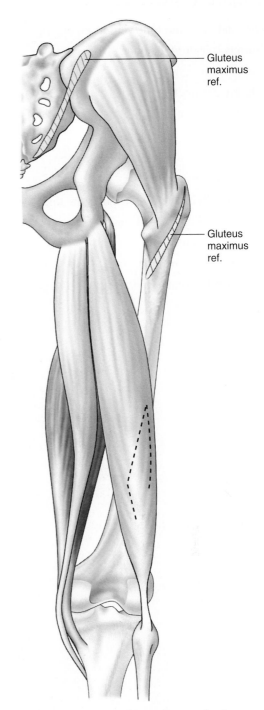

FIG. 10.24 Superficial muscles acting on the leg, posterior view.

Deep Muscles Acting on the Thigh, Posterior Views

Locate and *label* the following in Figure 10.25.

piriformis (pir-ih-FOR-mis)

    o: sacrum

    i: greater trochanter

    a: laterally rotates and abducts the thigh

obturator (OB-too-ra-tor) internus

    o: inner edges of obturator foramen

    i: greater trochanter

    a: laterally rotates and abducts the thigh

superior gemellus (jem-EL-lus)

    o: ischial spine

    i: greater trochanter

    a: laterally rotates and abducts the thigh

inferior gemellus

    o: ischial tuberosity

    i: greater trochanter

    a: laterally rotates and abducts the thigh

quadratus femoris

    o: ischial tuberosity

    i: proximal, posterior femur

    a: laterally rotates and abducts the thigh

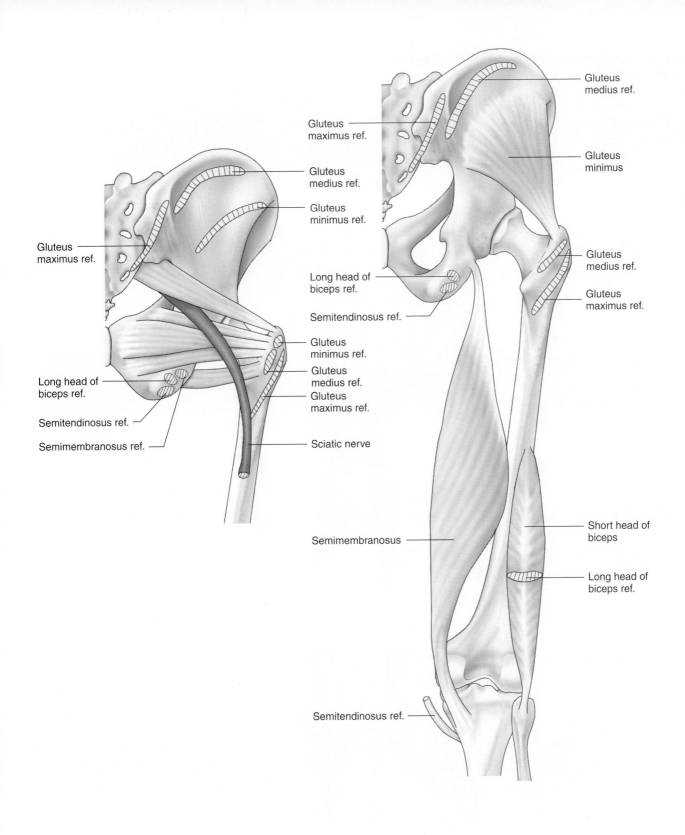

FIG. 10.25 Deep muscles acting on the thigh, posterior views.

MUSCLES OF THE INFERIOR APPENDAGE

Locate and *label* the muscles of the thigh and buttocks in Figure 10.26. *Use* the diagram to *understand* the relative positions of the muscles.

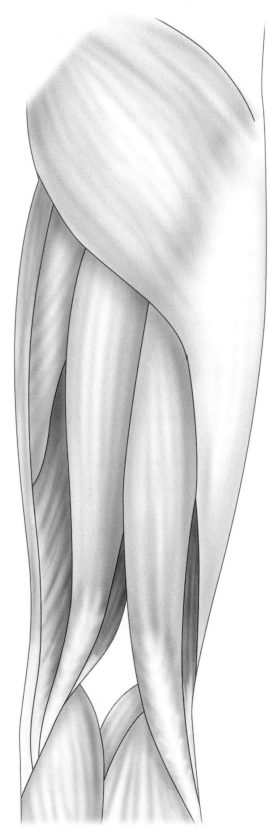

FIG. 10.26 Superficial muscles of the thigh and buttocks, posterior view.

CRITICAL THINKING

1. Which is NOT true of the hamstrings?

   a. They are extensors of the thigh.
   b. They are flexors of the leg.
   c. They include the semimembranosus.
   d. All parts originate on the ischial tuberosity.

2. Which is NOT an adductor of the brachium?

   a. pectoralis major
   b. coracobrachialis
   c. latissimus dorsi
   d. supraspinatus

3. The trapezius muscle is named by its

   a. shape.
   b. location.
   c. size.
   d. action.

4. The process of turning the palm anterior is

   a. pronation.
   b. medial rotation.
   c. adduction.
   d. supination.

5. The muscles that contract when clenching the hand into a fist are

   a. on the back of the brachium.
   b. all on the palm of the hand.
   c. in the wrist.
   d. on the anterior forearm.

6. List the muscles groups, in order of use, necessary to reach behind the back and scratch the heel while remaining in a standing position. Begin in anatomical position.

   _____

   _____

   _____

7. Provide the meanings of the following muscle names.

   gluteus medius _____

   _____

   splenius capitis _____

   _____

   intercostalis externus _____

   _____

   rectus abdominis _____

   _____

   piriformis _____

   _____

Deep Muscles Acting on the Medial Thigh, Anterior Views

Locate and *label* the following in Figure 10.27.

pectineus (pek-TIN-e-us)

> o: pubis
>
> i: proximal femur
>
> a: adducts the thigh

adductor longus

> o: pubis near the symphysis
>
> i: linea aspera
>
> a: adducts, flexes, and medially rotates the thigh

adductor brevis

> o: inferior portion of the pubis near juncture with ischium
>
> i: proximal linea aspera
>
> a: adducts, flexes, and medially rotates the thigh

adductor magnus

> o: inferior portion of the pubis and ischium
>
> i: linea aspera
>
> a: anterior portion—adducts, medially rotates, and flexes the thigh
> posterior portion—adducts, medially rotates, and extends the thigh

iliopsoas (il-e-o-SO-as)

> psoas (SO-as) major
>
>> o: transverse processes of the lumbar vertebrae
>>
>> i: lesser trochanter
>>
>> a: flexes the thigh and vertebral column
>
> iliacus (il-e-ak-us)
>
>> o: iliac fossa
>>
>> i: tendon of psoas major
>>
>> a: flexes the thigh

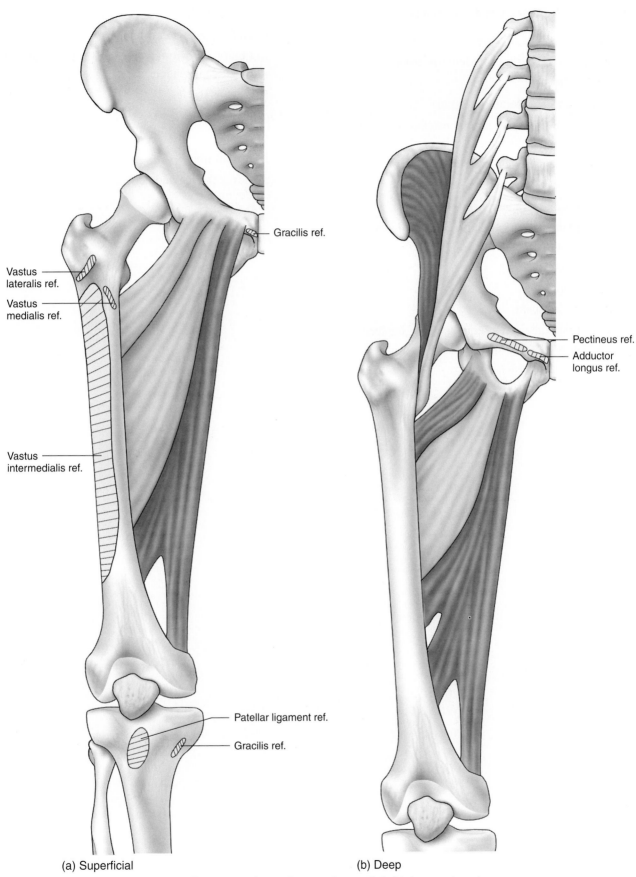

(a) Superficial (b) Deep

FIG. 10.27 Deep muscles acting on the medial thigh, anterior views.

Muscles Acting on the Anterior Leg

Locate and *label* the following in Figure 10.28.

quadriceps (KWOD-rih-ceps) femoris group

    *rectus femoris

        o: anterior, inferior iliac spine

        i: tibial tuberosity via the patellar ligament and quadriceps tendon

        a: extends the leg and flexes the thigh

    *vastus (VAS-tus) lateralis

        o: linea aspera and greater trochanter

        i: tibial tuberosity via the patellar ligament and quadriceps tendon

        a: extends the leg

    *vastus medialis

        o: linea aspera

        i: tibial tuberosity via the patellar ligament and quadriceps tendon

        a: extends the leg

vastus intermedius

    o: anterior and lateral femur

    i: tibial tuberosity via the patellar ligament and quadriceps tendon

    a: extends the leg

*gracilis (gras-IL-is)

    o: pubis

    i: medial tibia

    a: flexes the leg and adducts the thigh

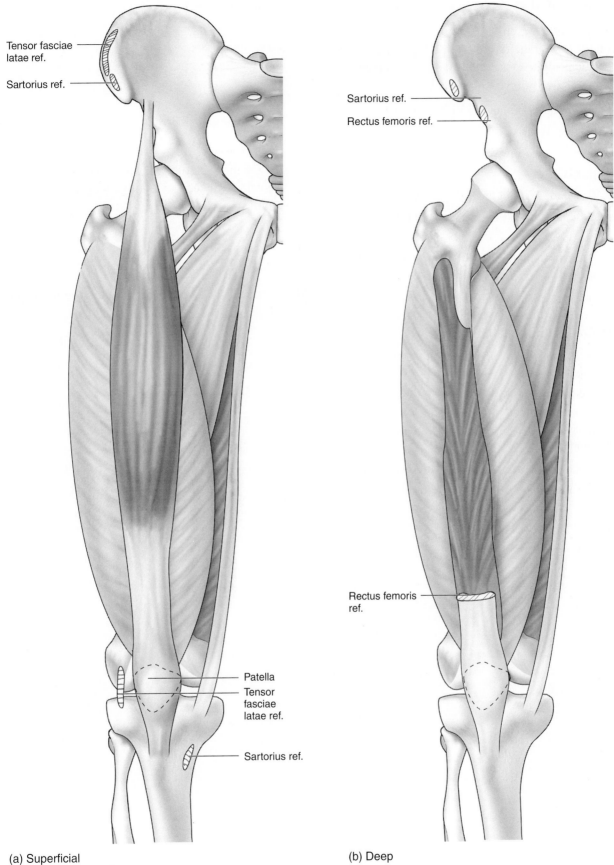

Tensor fasciae latae ref.

Sartorius ref.

Patella

Tensor fasciae latae ref.

Sartorius ref.

Sartorius ref.

Rectus femoris ref.

Rectus femoris ref.

(a) Superficial

(b) Deep

FIG. 10.28 Muscles acting on the anterior leg.

Muscles of the Anterior Thigh

Locate and *label* the muscles of the anterior thigh in Figure 10.29. *Use* the diagram to *understand* the relative positions of the muscles. In addition, *find* the muscles in the list below that act on the leg.

Locate and *label* the following in Figure 10.29.

*sartorius (sar-TO-re-us)

    o: anterior superior iliac spine

    i: medial tibia

    a: flexes the leg and thigh

*tensor fasciae latae (TEN-sor FA-she-e LA-te)

    o: iliac crest

    i: tibia through fascia of lateral thigh (iliotibial tract)

    a: flexes and abducts the thigh

CRITICAL THINKING

1. Which of the following inserts on the tibia via the patellar ligament?

    a. rectus femoris
    b. coracobrachialis
    c. latissimus dorsi
    d. supraspinatus

2. Provide the meanings of the following muscle names.

    gracilis _____

    _____

    sartorius _____

    _____

quadriceps femoris _____

_____

psoas major _____

_____

vastus lateralis _____

_____

3. On the back of the knee is a depression. What tendons form the borders of this popliteal fossa?

_____

_____

_____

Match the muscle in column A with its synergist in column B.

A	B
_____ **4.** deltoideus	a. quadriceps femoris
_____ **5.** iliocostalis cervicis	b. brachialis
_____ **6.** adductor magnus	c. supraspinatus
_____ **7.** biceps brachii	d. longissimus cervicis
_____ **8.** piriformis	e. gracilis
_____ **9.** rectus femoris	f. vastus medialis

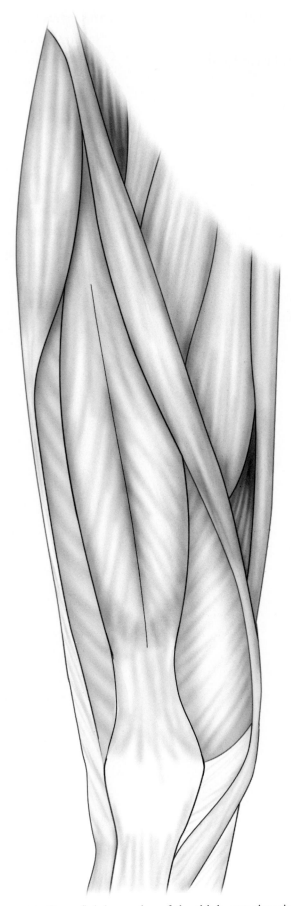

FIG. 10.29 Superficial muscles of the thigh, anterior view.

Muscles Acting on the Foot, Anterior View

Locate and *label* the following in Figure 10.30.

*tibialis anterior

    o: lateral condyle of the tibia

    i: first metatarsal and first cuneiform

    a: plantar extension (dorsiflexion)

extensor hallucis (HAL-luh-cis) longus

    o: fibula

    i: distal phalanx of hallux

    a: extends hallux and plantar extension

extensor digitorum longus

    o: lateral condyle of tibia and anterior fibula

    i: middle and distal phalanges of digits 2–5

    a: extends digits 2–5 and plantar extension

peroneus tertius (per-o-NE-us TER-shus)

    o: interosseus membrane and distal fibula

    i: fifth metatarsal

    a: plantar extension and everts the foot

peroneus brevis

    o: fibula

    i: fifth metatarsal

    a: plantar flexion and everts the foot

peroneus longus

    o: lateral condyle of tibia and shaft of fibula

    i: first metatarsal and first cuneiform

    a: plantar flexion and everts the foot

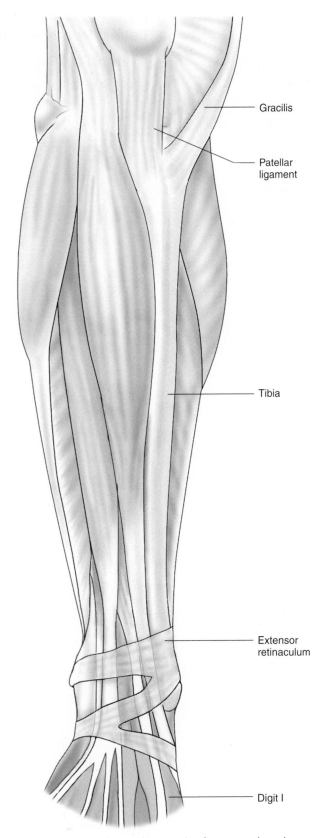

Gracilis

Patellar ligament

Tibia

Extensor retinaculum

Digit I

FIG. 10.30 Muscles acting on the foot, anterior view.

MUSCLES OF THE INFERIOR APPENDAGE

141

Muscles Acting on the Foot, Superficial, Posterior View

Locate and *label* the following in Figure 10.31(a).

*gastrocnemius (gas-tr<u>o</u>k-N<u>E</u>-m<u>e</u>-us)

    o: distal femur

    i: calcaneus via calcaneal (kal-K<u>A</u>-n<u>e</u>-al) tendon

    a: plantar flexion and flexes the leg

plantaris

    o: superior to lateral condyle of femur

    i: calcaneal tendon

    a: planter flexion and flexes the leg

Muscles Acting on the Foot, Middle Layer, Posterior View

Locate and *label* the following in Figure 10.31(b).

soleus (S<u>O</u>-l<u>e</u>-us)

    o: proximal tibia and fibula

    i: calcaneus via calcaneal tendon

    a: plantar flexion

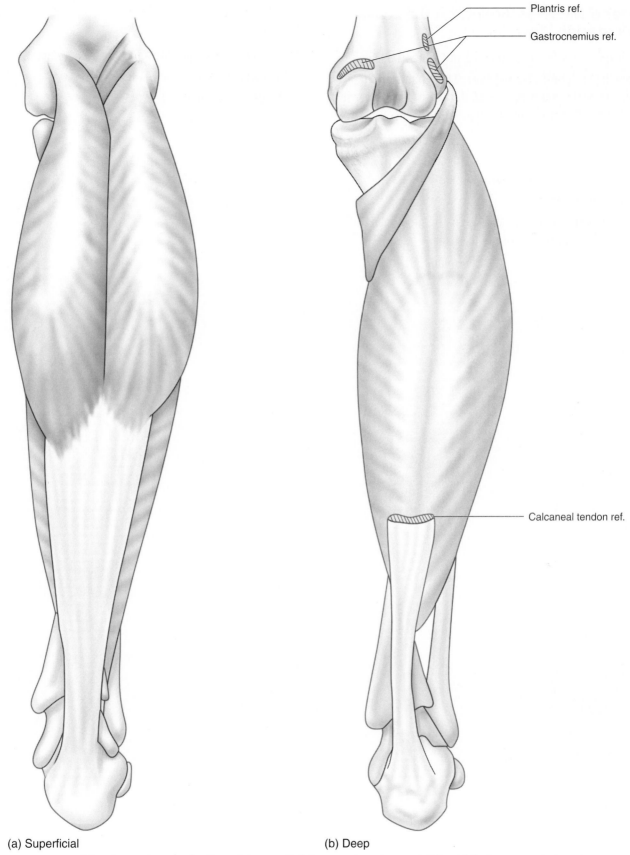

Plantris ref.

Gastrocnemius ref.

Calcaneal tendon ref.

(a) Superficial   (b) Deep

FIG. 10.31 Muscles acting on the foot: (a) superficial, posterior view and (b) middle layer, posterior view.

MUSCLES OF THE INFERIOR APPENDAGE   **143**

Deep Muscles Acting on the Foot, Posterior View

Locate and *label* the following in Figure 10.32.

popliteus ([pop-LIT-e̱-us] *Note:* Though the popliteus is in the same compartment of the leg as the following muscles, it acts on the leg, not on the foot.)

    o: lateral condyle of the femur

    i: proximal tibia

    a: flexion and medial rotation of the leg

flexor hallucis longus

    o: distal fibula

    i: distal phalanx of the hallux

    a: flexion of the hallux and plantar flexion

tibialis posterior

    o: interosseus membrane

    i: second through fourth metatarsal and tarsals

    a: inverts the foot and plantar flexion

flexor digitorum longus

    o: posterior tibia

    i: distal phalanges of digits 2–5

    a: flexes digits 2–5 and plantar flexion

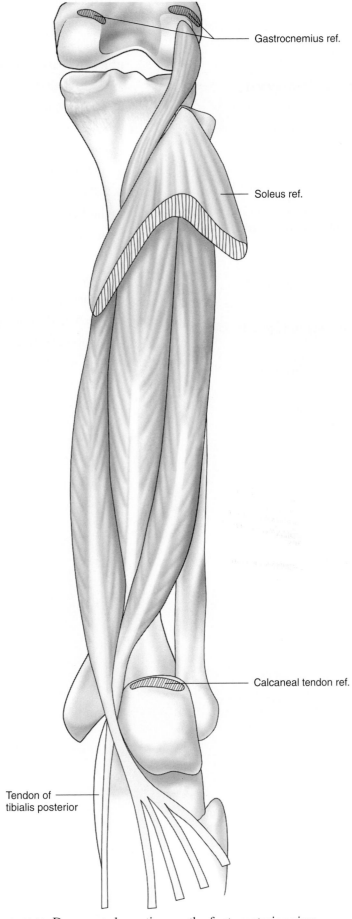

Gastrocnemius ref.

Soleus ref.

Calcaneal tendon ref.

Tendon of
tibialis posterior

FIG. 10.32 Deep muscles acting on the foot, posterior view.

Intrinsic Muscles of the Foot, Dorsal View

Locate and *label* the following in Figure 10.33(a).

extensor digitorum brevis

> o: calcaneus and the inferior portion of the extensor retinaculum
>
> i: extensor digitorum longus tendons of digits 2–4 and the proximal phalanx of the pollex
>
> a: extends digits 1–4

Intrinsic Muscles of the Foot, Plantar View

Locate and *label* the following in Figure 10.33(b).

abductor hallucis

> o: calcaneus and plantar aponeurosis (ap-o-noo-RO-sis)
>
> i: outside of proximal phalanx of hallux
>
> a: abducts and flexes the hallux

flexor digitorum brevis

> o: calcaneus and plantar aponeurosis
>
> i: middle phalanx of digits 2–5
>
> a: flexes digits 2–5

abductor digiti minimi

> o: calcaneus and plantar aponeurosis
>
> i: outside of proximal phalanx of digit 5
>
> a: abducts and flexes digit 5

flexor hallucis brevis

> o: third cuneiform and cuboid
>
> i: proximal phalanx of hallux
>
> a: flexes hallux

adductor hallucis

> o: second metatarsal through fourth
>
> i: inside of proximal phalanx of hallux
>
> a: adducts and flexes hallux

flexor digiti minimi brevis

> o: fifth metatarsal
>
> i: outside of proximal phalanx of digit 5
>
> a: flexes digit 5

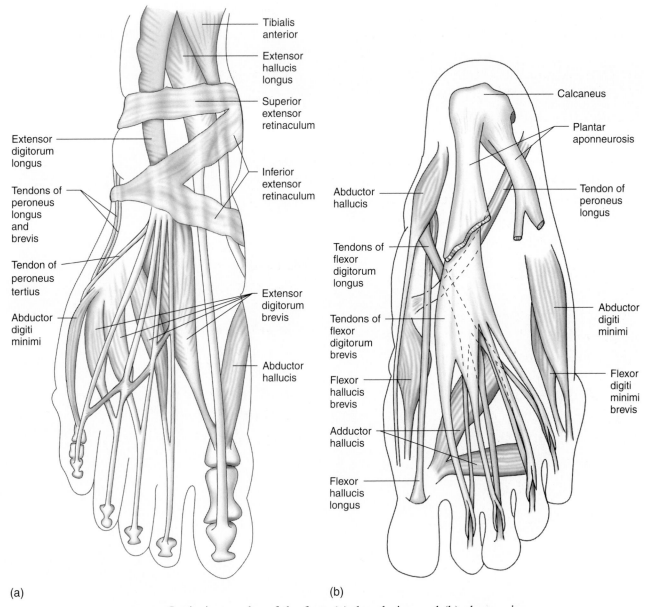

Tibialis anterior

Extensor hallucis longus

Superior extensor retinaculum

Extensor digitorum longus

Inferior extensor retinaculum

Tendons of peroneus longus and brevis

Tendon of peroneus tertius

Abductor digiti minimi

Extensor digitorum brevis

Abductor hallucis

Calcaneus

Plantar aponneurosis

Abductor hallucis

Tendon of peroneus longus

Tendons of flexor digitorum longus

Tendons of flexor digitorum brevis

Abductor digiti minimi

Flexor hallucis brevis

Flexor digiti minimi brevis

Adductor hallucis

Flexor hallucis longus

(a)

(b)

FIG. 10.33 Intrinsic muscles of the foot: (a) dorsal view and (b) plantar view.

SHORT ANSWER

1. *List* the muscles that

   elevate the hyoid

   compress the abdomen

   flex the spine

   extend the thigh

   flex the leg

   extend (dorsiflex) the foot

pronate the hand

2. *List* in order the muscle groups you would use in rising up from a squatting position to reach a book on a high shelf.

MATCHING

Match the muscle in column A with the action in column B. (Some actions may be used more than once.)

A	B
_____ 1. soleus	a. abduction
_____ 2. frontalis	b. adduction
_____ 3. teres minor	c. compression
_____ 4. pectoralis major	d. depression
_____ 5. rhomboideus	e. elevation
_____ 6. temporalis	f. extension
_____ 7. spinalis	g. flexion
_____ 8. sternocleidomastoideus	h. lateral rotation
_____ 9. digastricus	i. medial rotation
_____ 10. teres major	j. retraction
_____ 11. transverse abdominis	k. protraction
_____ 12. iliacus	
_____ 13. deltoideus	
_____ 14. palmaris longus	
_____ 15. tibialis anterior	
_____ 16. geniohyoideus	
_____ 17. stylohyoid (not elevation)	

Match the muscle in column A with the synergist in column B.

A

_____ 1. masseter

_____ 2. iliacus

_____ 3. semispinalis capitis

_____ 4. trapezius

_____ 5. stylohyoideus

_____ 6. pectineus

_____ 7. latissimus dorsi

_____ 8. teres minor

_____ 9. peroneus brevis

_____ 10. brachioradialis

_____ 11. palmaris longus

B

a. soleus

b. digastricus

c. infraspinatus

d. medial pterygoid

e. pectoralis major

f. rectus femoris

g. rhomboideus

h. splenius capitis

i. adductor longus

j. flexor carpi ulnaris

k. biceps brachii

Match the muscle in column A with the antagonist in column B.

A

_____ 1. zygomaticus major

_____ 2. sternohyoideus

_____ 3. frontalis

_____ 4. hamstring

_____ 5. erector spinae

_____ 6. supraspinatus

_____ 7. extensor carpi radialis longus

_____ 8. gluteus minimus

_____ 9. triceps brachii

_____ 10. tibialis anterior

_____ 11. rhomboideus major

B

a. brachialis

b. coracobrachialis

c. depressor anguli oris

d. flexor carpi ulnaris

e. quadriceps femoris

f. procerus

g. rectus abdominis

h. mylohyoideus

i. adductor brevis

j. serratus anterior

k. peroneus brevis

Match the muscles in column A with the insertions in column B.

	A			B
_____	1. spinalis thoracis		a.	first metacarpal
_____	2. sternothyroideus		b.	fascia of lateral thigh
_____	3. diaphragm		c.	medial tibia
_____	4. orbicularis oculi		d.	pinna
_____	5. tensor fasciae latae		e.	proximal head of fibula
_____	6. levator scapulae		f.	skin around eye
_____	7. quadratus lumborum		g.	skull
_____	8. biceps femoris		h.	spinous processes of superior thoracic vertebrae
_____	9. any capitis muscle		i.	thyroid cartilage of larynx
_____	10. sartorius		j.	rib 12 and transverse processes of L_1–L_4
_____	11. auricularis anterior		k.	vertebral border of scapula
_____	12. abductor pollicis longus		l.	central tendon
_____	13. tibialis posterior		m.	ribs
_____	14. intercostalis		n.	second through fourth metatarsal, tarsals

Match the muscle in column A with the origins in column B.

	A			B
_____	1. temporalis		a.	anterior, inferior iliac spine
_____	2. sternocleidomastoideus		b.	apex of mandible
_____	3. iliacus		c.	clavicle, sternum
_____	4. digastricus		d.	coracoid process of scapula
_____	5. coracobrachialis		e.	ilium, ribs, vertebrae
_____	6. pectoralis minor		f.	iliac fossa
_____	7. adductor magnus		g.	mastoid process and mandible
_____	8. rectus femoris		h.	pubis
_____	9. geniohyoideus		i.	ribs 3–5
_____	10. transverse abdominis		j.	temporal bone
_____	11. extensor digitorum longus		k.	lateral condyle of tibia, fibula

Match the muscle in column A with the antagonist in column B.

A

_____ 1. flexor hallucis longus

_____ 2. soleus

_____ 3. adductor hallucis

_____ 4. gracilis

_____ 5. popliteus

_____ 6. extensor digitorum longus

B

a. gluteus minimus

b. abductor hallucis

c. flexor digiti minimi

d. tibialis anterior

e. vastus intermedius

f. extensor hallucis longus

Match the muscle in column A with the origin in column B.

A

_____ 1. deltoid

_____ 2. adductor brevis

_____ 3. depressor anguli oris

_____ 4. sternohyoideus

_____ 5. abductor digiti minimi

_____ 6. longissimus cervicis

B

a. inferior, posterior portion of pubis

b. calcaneus

c. transverse processes of T_4, T_5

d. clavicle and acromion process

e. mandible

f. clavicle and manubrium

Skeletal Muscles of the Cat

Contents

Objectives

1. Dissect the cat muscular system to better understand that of humans.

2. Understand the subtle structural relationships that do not appear on plastic models.

BLUNT DISSECTION

The blunt dissection (dih-SEK-shun) technique will be used as you study the cat. Use the scissors as little as possible. No scalpel is required. The techniques of blunt dissection preserve the specimen and result in a superior dissection.

During dissection, muscles are **isolated** (separated from surrounding tissues) using the probe, fingers, and forceps. Your instructor will demonstrate the techniques required. Once a muscle is isolated, it may be **reflected** (cut midway between the origin and insertion and then turned back). In this way, the ends remain attached for

later study. You should *never* completely remove a muscle or other structure unless so instructed.

SKINNING THE SPECIMEN

⚠ THERE IS A RISK OF SPLASHING. WEAR SAFETY GOGGLES. YOU MAY ALSO WISH TO WEAR GLOVES TO REDUCE SKIN IRRITATION.

After *removing* the specimen from the shipping bag, *drain* the bag in the sink and *rinse* the cat well under cool tap water before skinning. Take just a moment and *familiarize* yourself with the following landmarks while *comparing* them to the human.

dorsal (D<u>O</u>R-sal) related to the back

ventral (VEN-tral) related to the belly

cranial (KR<u>A</u>-n<u>e</u>-al) related to the head

caudal (KAW-dl) related to the tail

Determine the sex of your specimen. *Refer* to Figure 11.1.

⚠ ALL OF THE FOLLOWING CUTS ARE MADE THROUGH THE SKIN ONLY. DO NOT CUT THE UNDERLYING MUSCLES.

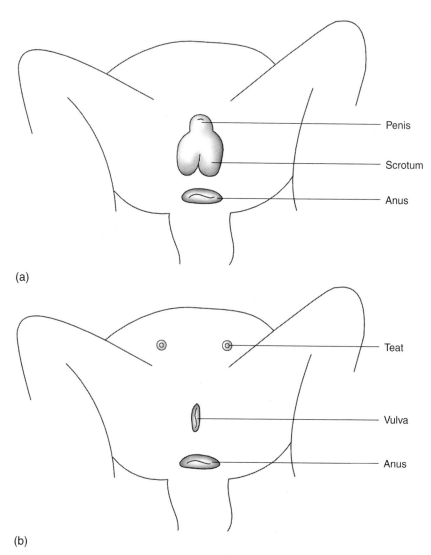

FIG. 11.1 External genitalia, (a) male and (b) female.

1. *Make* a midventral incision from the apex of the mandible to the pubic symphysis.

2. *Make* an incision from the first cut down each appendage to the wrist and ankle.

3. *Make* a circular incision around each wrist and ankle connecting to the previous cut.

4. *Make* an incision along the mandible, across the cheek, caudal to the eye on each side and join the two cuts on the forehead.

5. *Make* a circular incision around each ear, at the base of the tail, and around the external genitalia.

6. *Using* your probe, fingers, and occasionally your scissors, remove the skin.

    a. *Avoid* cutting underlying muscle.

    b. *Do not damage* the external genitalia.

    c. *Ask* your instructor if your specimen is triple injected. If so, *make* a circular cut around the sutures on the right side of the abdomen.

CLEANING UP

Unless you are directed otherwise, *wrap* the skin in paper towels and *discard* it in the proper receptacle. DO NOT RINSE the specimen after it is skinned. When you finish a dissection laboratory exercise, always *do* the following:

1. *Tie* a label, with your name printed in pencil, onto the leg of the specimen.

2. *Place* your specimen in the plastic storage bag provided, and exclude as much air as possible.

3. *Tie* the bag shut with a string, rubber band, or twist tie.

4. *Place* the specimen in the designated storage area.

5. *Empty* your dissection tray into the waste bin, *wipe* it clean, and *rinse* it under the tap.

⚠ *DO NOT ALLOW* MATERIALS TO ENTER THE DRAIN.

6. *Stack* your tray, inverted, on the drainboard of the sink or other designated area.

Some cat muscles have been omitted from the diagrams presented here to promote clarity and brevity. Only those muscles important to the understanding of human musculature are included. You should do the following:

1. *Follow* the dissection instructions provided in brackets after the muscle name.

2. *Consult* the diagrams for help in locating the muscles.

3. *Learn* the human muscle name if the cat muscle name differs.

4. *Understand* that the objective of this chapter is to become familiar with the muscles of the human body, not the cat. Some of the separate muscles found in the cat are fused into single muscles in the human. A few of the muscles present in cats are absent in humans. Where it is necessary to understanding the musculature of the human, the fusions and deletions will be noted.

MUSCLES OF THE HEAD AND NECK

Superficial Muscles of the Head and Neck, Ventral View

Locate the following muscles on the cat. Consult Figure 11.2(a) and CDA-1.

masseter

digastricus

mylohyoideus [*isolate* and *reflect*]

sternohyoideus [*isolate* and *reflect*]

sternomastoideus (sternocleidomastoideus in humans)

temporalis (not illustrated)

Deep Muscles of the Head and Neck, Ventral View

Locate the following muscles on the cat. Consult Figure 11.2(b) and CDA-1.

geniohyoideus

sternothyroideus

thyrohyoideus

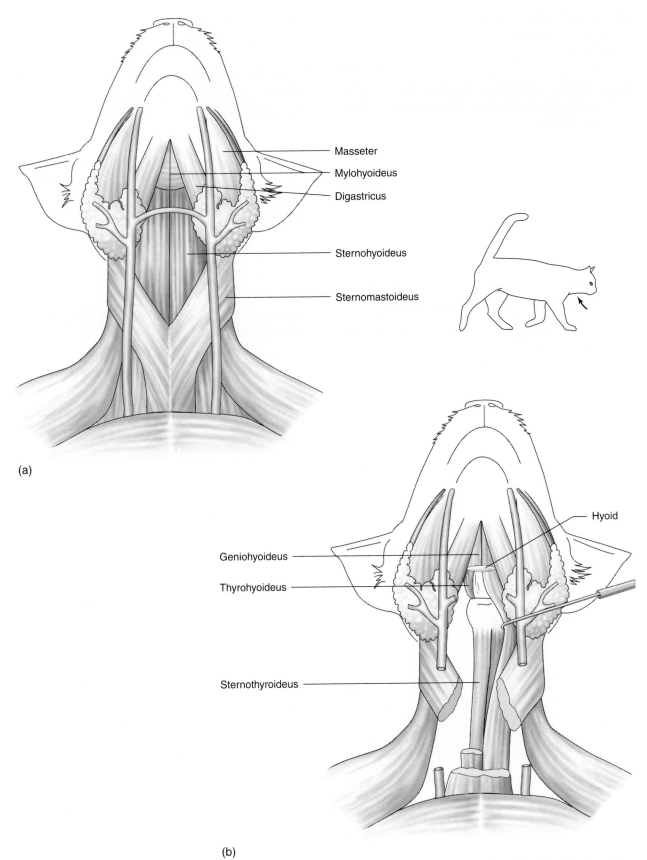

Masseter

Mylohyoideus

Digastricus

Sternohyoideus

Sternomastoideus

(a)

Geniohyoideus

Thyrohyoideus

Hyoid

Sternothyroideus

(b)

FIG. 11.2 (a) Superficial muscles of the head and neck, ventral view, and (b) deep muscles of the head and neck, ventral view.

MUSCLES ACTING ON THE SCAPULA AND BRACHIUM, VENTRAL

Superficial Muscles Acting on the Scapula and the Brachium, Ventral View

Locate the following muscles on the cat. Consult Figure 11.3(a) and CDA-2.

pectoroantebrachialis (absent in humans)

pectoralis major [*isolate* and *reflect*]

pectoralis minor [*isolate* and *reflect*]

xiphihumeralis (absent in humans) [*isolate* and *reflect*]

Deep Muscles Acting on the Scapula and the Brachium, Ventral View

Locate the following muscles on the cat. Consult Figure 11.3(b) and CDA-3.

serratus ventralis (serratus anterior in humans)

subscapularis [*pull* the scapulae laterad]

MUSCLES OF THE THORAX AND THE ABDOMEN

Locate the following muscles in Figure 11.3(b) and CDA-3.

transversus costalis (absent in humans)

scalenus (present in the humans but not illustrated)

rectus abdominis

external oblique [*isolate* and *reflect*] (this muscle is very thin)

internal oblique [*isolate* and *reflect*] (this muscle is very thin)

transversus abdominis (often adhering to the deep side of internal oblique)

external intercostalis [*isolate* and *reflect*]

internal intercostalis (not illustrated, the fibers run obliquely to external)

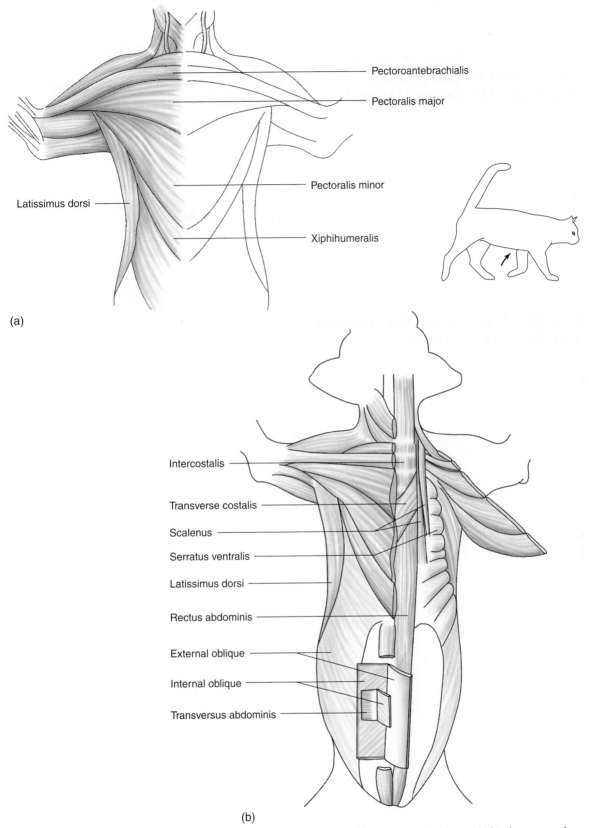

FIG. 11.3 (a) Superficial muscles acting on the scapula and the brachium, ventral view and (b) deep muscles acting on the scapula and the brachium and muscles of the thorax and the abdomen, ventral view.

MUSCLES ACTING ON THE SCAPULA AND BRACHIUM, DORSAL

Muscles Acting on the Scapula and the Brachium, Dorsal and Lateral

Locate the following muscles on the cat. Consult Figure 11.4(a) and CDA-5.

clavotrapezius ⎤
acromiotrapezius ⎟ (trapezius in humans) [*isolate* and
spinotrapezius ⎦ *reflect*]

acromiodeltoid ⎤
spinodeltoid ⎟ (deltoid in humans)
clavodeltoid ⎦

latissimus dorsi [*isolate* and *reflect*]

Deep Muscles Acting on the Scapula and the Brachium, Dorsal

Locate the following muscles on the cat. Consult Figure 11.4(b) and CDA-6.

rhomboideus capitis (rhomboideus minor in humans) [*isolate* and *reflect*]

rhomboideus (rhomboideus major in humans) [*isolate* and *reflect*]

levator scapulae ventralis (levator scapulae in humans)

supraspinatus

infraspinatus

teres major

MUSCLES ACTING ON THE SPINE

Locate the following muscles on the cat. Consult Figure 11.4(b) and CDA-4.

[*Peel* away the fascia in the lumbar region to expose the following muscles.]

spinalis

longissimus

iliocostalis

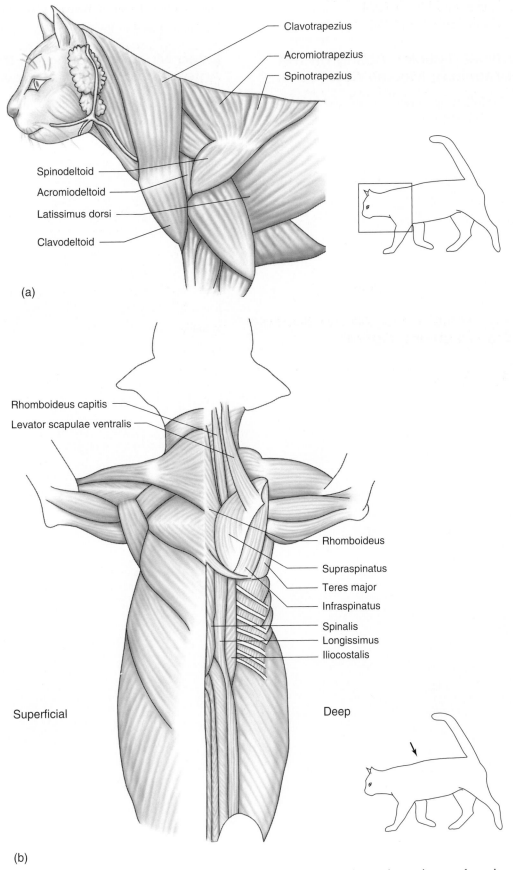

Clavotrapezius

Acromiotrapezius

Spinotrapezius

Spinodeltoid

Acromiodeltoid

Latissimus dorsi

Clavodeltoid

(a)

Rhomboideus capitis

Levator scapulae ventralis

Rhomboideus

Supraspinatus

Teres major

Infraspinatus

Spinalis

Longissimus

Iliocostalis

Superficial

Deep

(b)

FIG. 11.4 Superficial and deep muscles acting on the scapula and brachium and muscles acting on the spine, lateral (a) and dorsal (b) views.

MUSCLES ACTING ON THE SPINE, DORSAL

MUSCLES ACTING ON THE ANTEBRACHIUM

Superficial Muscles Acting on the Antebrachium, Medial View

Locate the following muscles on the cat. Consult Figure 11.5(a) and CDA-7a.

pectoroantebrachialis (absent in humans) [*isolate* and *reflect*]

clavodeltoid (absent in humans) [*isolate* and *reflect*]

brachialis (deep to the two muscles listed above)

Deep Muscles Acting on the Antebrachium, Medial View

Locate the following muscles on the cat. Consult Figure 11.5(b) and CDA-7b.

triceps brachii

biceps brachii

subscapularis

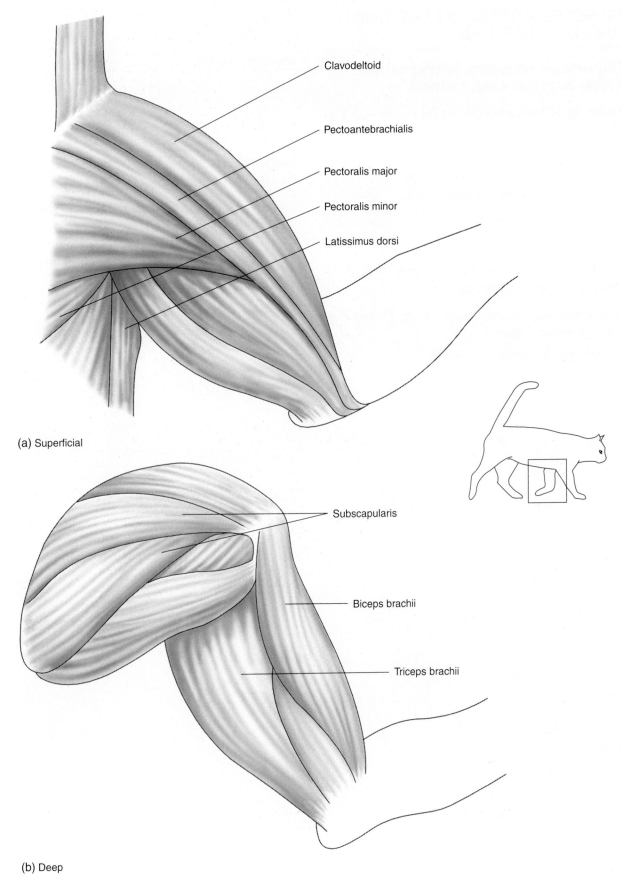

(a) Superficial

Clavodeltoid

Pectoantebrachialis

Pectoralis major

Pectoralis minor

Latissimus dorsi

Subscapularis

Biceps brachii

Triceps brachii

(b) Deep

FIG. 11.5 (a) Superficial and (b) deep muscles acting on the antebrachium, medial views.

MUSCLES ACTING ON THE THIGH AND LEG, LATERAL

Superficial Muscles Acting on the Thigh and the Leg, Lateral

Locate the following muscles on the cat. Consult Figure 11.6(a) and CDA-8.

sartorius [*isolate* and *reflect*]

tensor fasciae latae [*isolate* and *reflect*]

biceps femoris [*isolate* and *reflect*]

gluteus maximus

gluteus medius

Deep Muscles Acting on the Thigh and the Leg, Lateral

Locate the following muscles on the cat. Consult Figure 11.6(b) and CDA-9.

caudofemoralis (absent in humans)

vastus lateralis

semitendinosus

semimembranosus

MUSCLES ACTING ON THE FOOT

Superficial Muscles Acting on the Foot, Posterior

Locate the following muscles on the cat. Consult Figure 11.6(b) on CDA-9.

gastrocnemius [*isolate* and *reflect*]

soleus

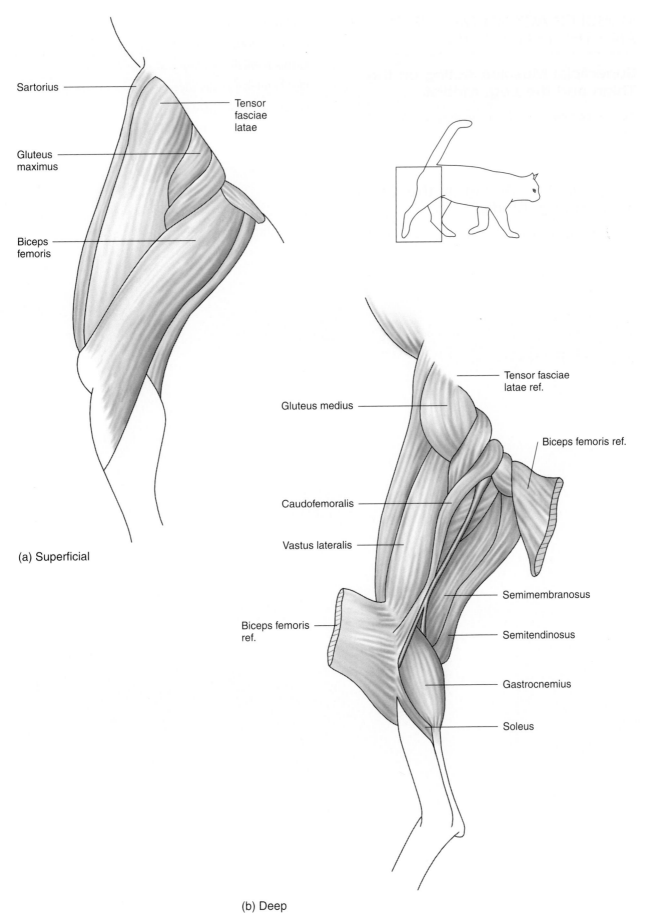

Sartorius

Tensor fasciae latae

Gluteus maximus

Biceps femoris

(a) Superficial

Gluteus medius

Tensor fasciae latae ref.

Biceps femoris ref.

Caudofemoralis

Vastus lateralis

Semimembranosus

Biceps femoris ref.

Semitendinosus

Gastrocnemius

Soleus

(b) Deep

FIG. 11.6 (a) Superficial and (b) deep muscles acting on the thigh, leg, and foot, lateral views.

MUSCLES ACTING ON THE FOOT

MUSCLES ACTING ON THE THIGH AND THE LEG, MEDIAL

Superficial Muscles Acting on the Thigh and the Leg, Medial

Locate the following muscle on the cat. Consult Figure 11.7(a) and CDA-10.

gracilis [*isolate* and *reflect*]

Deep Muscles Acting on the Thigh and the Leg, Medial

Locate the following muscles on the cat. Consult Figure 11.7(b) and CDA-11.

iliopsoas [*isolate*]

pectineus [*isolate*]

rectus femoris [*isolate* and *reflect*]

vastus medialis [*isolate* and *reflect*]

vastus intermedialis (deep to rectus femoris)

adductor longus [*isolate*]

adductor femoris (the combined adductor brevis and magnus of humans) [*isolate*]

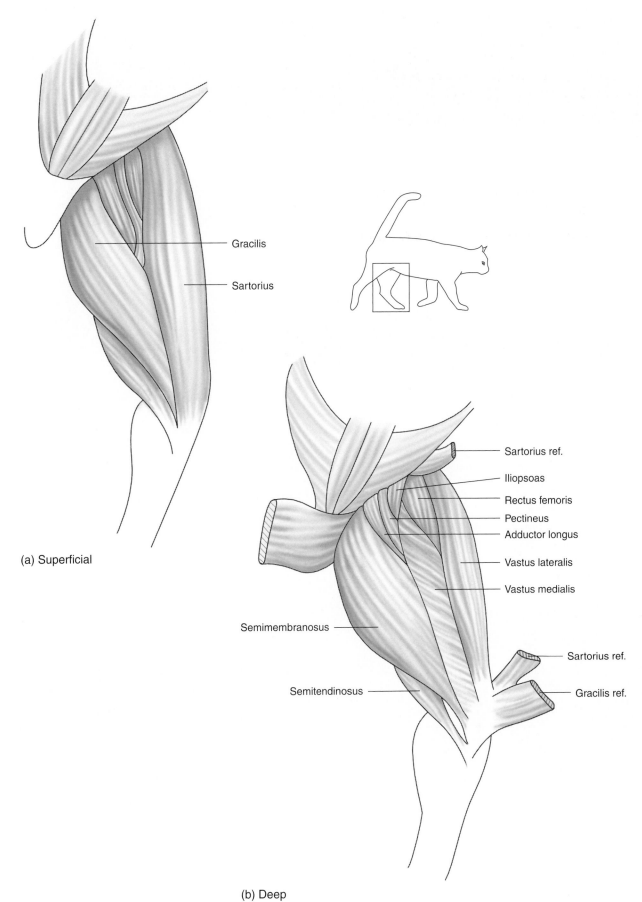

Gracilis

Sartorius

Sartorius ref.

Iliopsoas

Rectus femoris

Pectineus

Adductor longus

Vastus lateralis

Vastus medialis

Semimembranosus

Sartorius ref.

Semitendinosus

Gracilis ref.

(a) Superficial

(b) Deep

FIG. 11.7 (a) Superficial and (b) deep muscles acting on the thigh and leg, medial views.

12 Cardiovascular System

Contents

Objectives

1. List the components of the cardiovascular system.

2. Describe the types and locations of the major tissues of the cardiovascular system.

3. Describe the contents of the mediastinum.

4. List the chambers of the heart and the great vessels.

5. Describe the pulmonary loop and the systemic loop.

6. Name and describe each valve of the heart. Describe the function of each valve.

7. Describe the systemic vessels of the myocardium.

8. Describe the three types of blood vessels. List and describe each layer. Be able to identify the vessels and their components on a microscope slide.

9. Be able to locate on charts, models, and specimens, the major arteries and veins.

10. Define and describe the hepatic portal system.

11. List the blood solids and provide a function for each. Be able to identify each on a prepared blood slide.

COMPONENTS OF THE CARDIOVASCULAR SYSTEM

The cardiovascular system consists of the heart, blood vessels, and blood. This system functions to transport materials such as nutrients, wastes, hormones, gases, or cells that function to repair wounds and fight infection within the body. In this transportation system there is a pump (the heart), a pipeline (the vessels), and a fluid to carry the materials (the blood).

THE HEART

The human heart is a four-chambered pump with two receiving chambers (**atria**) and two pumping chambers (**ventricles**). The heart is located in the **mediastinum** posterior to the sternum. **Pericardial membranes** surround the heart. An outer **fibrous pericardium** forms a saclike structure and a two-layered **serous pericardium**

lies beneath. The serous pericardia are the **parietal pericardium,** adhering to the underside of the fibrous pericardium, and the **visceral pericardium** (epicardium), lying on the cardiac muscle (**myocardium**). Between the two serous pericardia is a presumptive space, the **pericardial cavity.** Deep to the myocardium and lining the chambers of the heart is the **endocardium.**

Mediastinum, Cross Section at Heart Level, Inferior View

Locate and *label* the following in Figure 12.1.

mediastinum (me-de-as-TI-num)

    fibrous pericardium (per-ih-KAR-de-um)

    serous (SE-rus) pericardium

        parietal (pa-RI-e-tal) pericardium

        visceral pericardium (epicardium [ep-ih-KAR-de-um])

    myocardium (mi-o-KAR-de-um)

    endocardium (en-do-KAR-de-um)

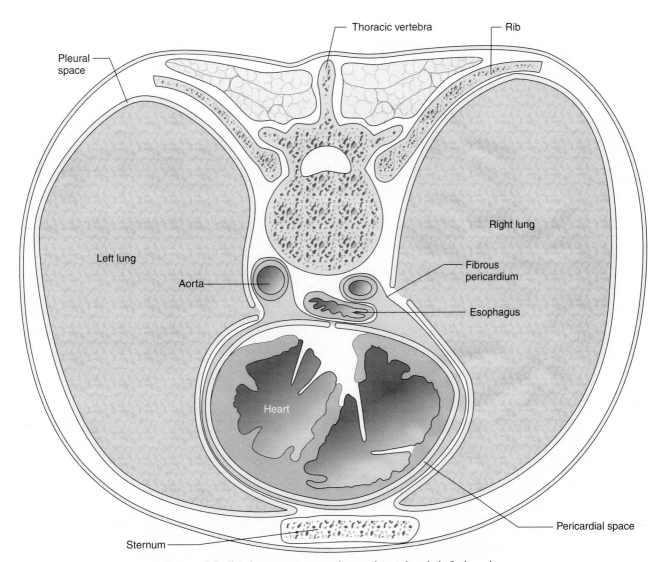

FIG. 12.1 Mediastinum, cross section at heart level, inferior view.

The vessels discussed in this chapter are listed below in outline form. For example, the indentations under arteries (see immediately below) indicate branches. That is, the brachiocephalic, left common carotid, and left subclavian arteries are branches of the aorta. For veins, a vessel indented below another indicates that the indented vessel is a tributary to the one listed above.

Heart, Anterior View, Superficial

Locate and *label* the following in Figure 12.2(a).

aorta (a̲-O̲R-tah)

    brachiocephalic (br̲ak-e-o̲-seh-FAL-ik) artery

    left common carotid (ka-ROT-id) artery

    left subclavian (sub-KLA̲-ve̲-an) artery

pulmonary (PUL-mo̲-ner-e̲) artery

superior vena cava (VE̲-nah KA̲-vah)

inferior vena cava

apex (A̲-peks)

Draw dashed lines on Figure 12.2(a) to illustrate the following, then *label* the diagram.

right atrium

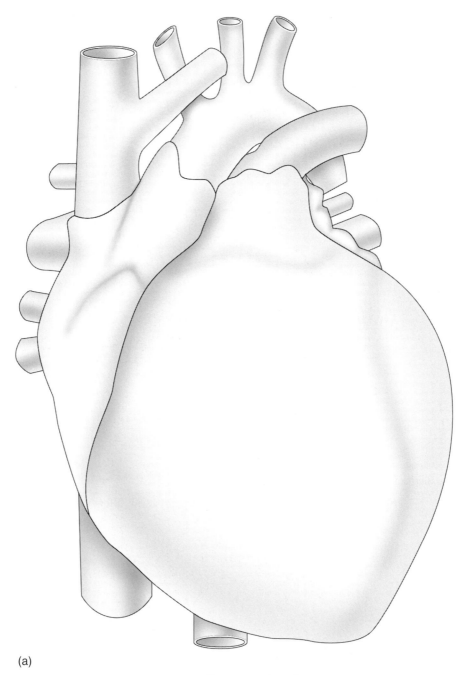

(a)

FIG. 12.2 (a) Heart, anterior view.

left atrium

right ventricle

left ventricle

Heart, Posterior View, Superficial

Locate and *label* the following in Figure 12.2(b).

aorta

    brachiocephalic artery

left common carotid artery

left subclavian artery

pulmonary artery

pulmonary vein

superior vena cava

inferior vena cava

apex

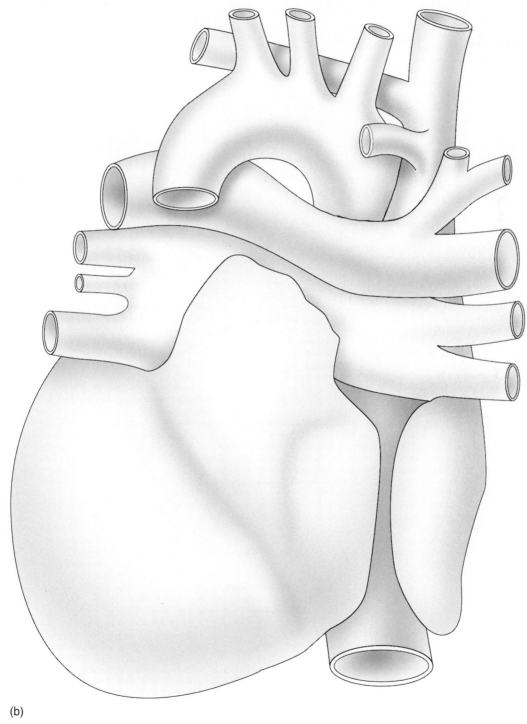

(b)

FIG. 12.2 (b) Heart, posterior view.

Heart, Frontal Section, Anterior View

Locate and *label* the following in Figure 12.3(a).

right atrium (<u>A</u>-tr<u>e</u>-um)

left atrium

right ventricle (VEN-tri-kl)

left ventricle

right atrioventricular (<u>a</u>-tr<u>e</u>-<u>o</u>-ven-TRIK-<u>u</u>-lar) or tricuspid (tr<u>i</u>-KUS-pid) valve

left atrioventricular, mitral (M<u>I</u>-tral), or bicuspid (b<u>i</u>-KUS-pid) valve

pulmonary semilunar (sem-<u>e</u>-LOO-nar) valve

aortic semilunar valve

chordae tendineae (K<u>O</u>R-d<u>e</u> TEN-dih-n<u>e</u>)

papillary (PAP-ih-l<u>ar</u>-<u>e</u>) muscle

sinoatrial (s<u>i</u>-n<u>o</u>-<u>A</u>-tr<u>e</u>-al) (SA) node

atrioventricular (AV) node

interventricular bundle or bundle of His

    left bundle branch

    right bundle branch

conduction myofibrils (m<u>i</u>-<u>o</u>-F<u>I</u>-bril) or Purkinje (pur-KIN-j<u>e</u>) fibers

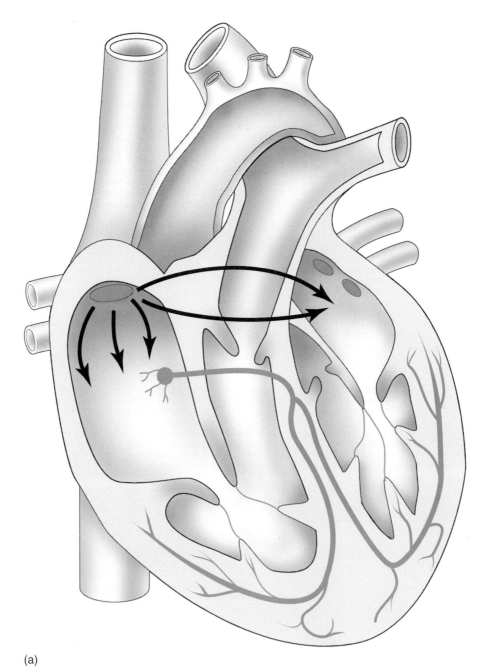

(a)

FIG. 12.3 (a) Heart, frontal section, anterior views, conduction system (a) and blood flow (b).

1. How does the conductive system of the heart function to control the cardiac cycle? Use a separate sheet of paper for your answer.

2. Trace a drop of blood through the heart from the entrance at the vena cava to the exit at the aorta. List, in the order of occurrence, all vessels, chambers, and valves. Use a separate sheet of paper for your answer.

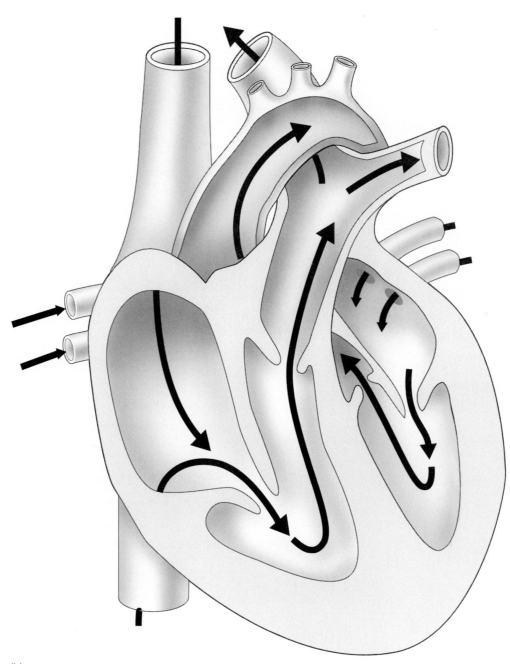

(b)

FIG. 12.3 (b) Illustrates the flow of blood through the heart. *Check* this pattern against your answer to question 2.

CORONARY CIRCULATION

The heart is muscle tissue and must have a good blood supply. The arteries that deliver blood to the heart tissue (**coronary circulation**) branch from the aorta just superior to the aortic semilunar valve.

Coronary Circulation, Anterior and Posterior Views of the Heart

After each of the vessels listed below, *describe* the area of the heart muscle supplied or drained and *label* Figure 12.4(a) and (b).

left coronary (K<u>O</u>R-o-n<u>a</u>-r<u>e</u>) artery

    anterior interventricular artery

    circumflex (SER-kum-fleks) artery

right coronary artery

    posterior interventricular artery

coronary sinus

    great cardiac vein

    middle cardiac vein

    small cardiac vein

CRITICAL THINKING

1. *Define:*

angina pectoris

_____

_____

_____

_____

myocardial infarction

_____

_____

_____

atherosclerosis

_____

_____

_____

dysrhythmia

_____

_____

_____

diastolic pressure (diastole)

_____

_____

_____

systolic pressure (systole)

_____

_____

_____

2. *List* at least five risk factors for heart disease.

_____

_____

_____

_____

_____

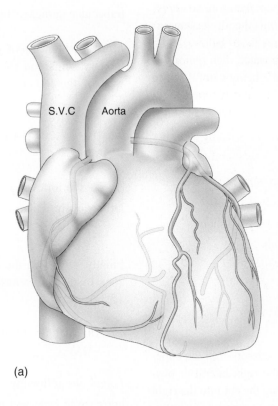

(a)

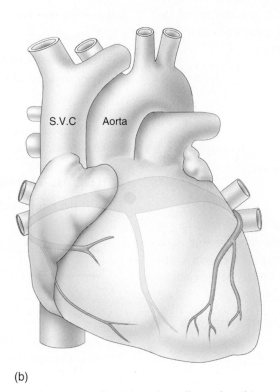

(b)

FIG. 12.4 Heart, coronary arteries (a) and cardiac veins (b), anterior views.

Name

VERTEBRATE HEART DISSECTION

Your instructor will provide a mammalian heart (beef, pig, or sheep). You should *follow* the dissection instructions and *locate* the structures listed below.

First, *rinse* the heart with tap water, then *remove* as much fatty tissue as possible while being careful not to damage any deeper tissues. While using Figure 12.5(a) and (b) as a guide, *locate* the following external features:

aorta

pulmonary artery

superior vena cava (very thin walls)

inferior vena cava (very thin walls)

pulmonary veins (these may be very short)

anterior interventricular sulcus

posterior interventricular sulcus

right and left auricles

right and left atria

right and left ventricles

Now *place* the blunt tip of your scissors into the superior vena cava and *cut* through the right lateral or posterior wall of the vessel, continuing the cut into the right atrium. *Stop* before entering the right ventricle (see dotted line 1 on Figure 12.5b). Pull open the cut so that you can see the top of the right atrioventricular valve and the interior of the right atrium.

Locate the following:

exit from the coronary sinus

exit from the inferior vena cava

pectinate (PEK-tih-nat) muscle

right atrioventricular or tricuspid valve

    3 cusps

Now *insert* the blunt tip of your scissors into the pulmonary trunk and *cut* through the anterior wall. Continue this incision through the wall of the right ventricle until it reaches the inferior boundary of the chamber (see dotted line 2 on Figure 12.5a).

⚠ TRY NOT TO MUTILATE THE RIGHT SEMILUNAR VALVE.

Locate the following:

right or pulmonary semilunar valve

    3 cusps

chordae tendineae of the tricuspid valve

papillary muscles of the tricuspid valve

trabeculae carneae (tra-BEK-u-le KAR-ne-e)

Insert the blunt tip of your scissors into one of the pulmonary veins and *cut* through the lateral or posterior wall of the left atrium (see dotted line 3 on Figure 12.5b).

⚠ DO NOT CUT INTO THE LEFT ATRIOVENTRICULAR VALVE.

Locate the following:

left atrioventricular, bicuspid, or mitral valve

    2 cusps

Finally, *cut* down the anterior wall of the aorta to the apex of the left ventricle (see dotted line 4 on Figure 12.5a).

⚠ TRY NOT TO MUTILATE THE LEFT SEMILUNAR VALVE.

Locate the following:

entrance to right and left coronary arteries

left (aortic) semilunar valve

    3 cusps

chordae tendineae of the bicuspid valve

papillary muscles of the bicuspid valve

When you finish, *ask* your instructor for a plastic bag in which to store the heart for later study and review. Make sure that you *tag* the specimen with your name.

CRITICAL THINKING

1. *List* the four valves of the heart and the function of each.

    _____

    _____

    _____

    _____

2. *Trace* a drop of blood. *Begin* and *finish* in the aorta. *Pass* through the anterior interventricular artery en route. *List* all vessels, chambers, and valves in the order of occurrence. *Use* a separate sheet of paper for your answer.

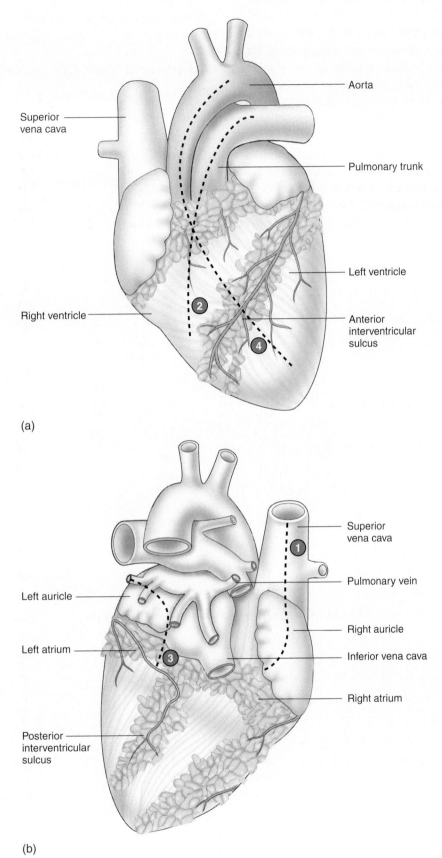

Aorta

Superior
vena cava

Pulmonary trunk

Left ventricle

Right ventricle

2

4

Anterior
interventricular
sulcus

(a)

Superior
vena cava

Pulmonary vein

Left auricle

Right auricle

Left atrium

Inferior vena cava

3

1

Right atrium

Posterior
interventricular
sulcus

(b)

FIG. 12.5 Sheep heart, anterior (a) and posterior (b) views.

ANATOMY OF THE BLOOD VESSEL

Arteries are vessels that carry blood **away** from the heart. They are more muscular and contain more elastic connective tissue in the walls than do veins or capillaries. **Veins,** which carry blood **toward** the heart, are less muscular with more inelastic (white fibrous) connective tissue than arteries. Also, the endothelial lining of veins below the heart is modified to form valves that maintain the flow of blood toward the heart. **Capillaries** are the sites of exchange between blood and tissues. They connect tiny arteries (arterioles) to tiny veins (venules). Capillaries consist of only one layer of endothelial cells and have an internal diameter of about 7 microns. *View* a microscope slide containing prepared cross sections of arteries and veins.

Artery and Vein Cross Sections

Locate and *label* the following in Figure 12.6(a) and (b).

tunica (TOO-nih-ka) externa

external elastic lamina (artery only)

tunica media

internal elastic lamina (artery only)

tunica interna

lumen (LOO-men)

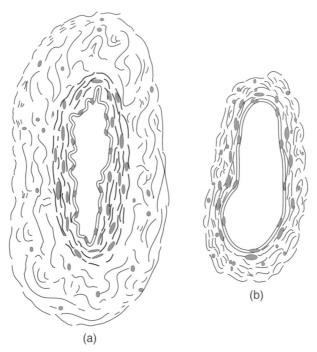

(a)

(b)

FIG. 12.6 Cross sections of artery (a) and vein (b).

CRITICAL THINKING

Define the following terms:

1. elastic artery

_____

_____

_____

_____

_____

2. muscular artery

_____

_____

_____

_____

_____

3. anastomosis

_____

_____

_____

_____

_____

4. sinusoid

_____

_____

_____

5. elastic lamina

_____

_____

_____

THE ARTERIES

Arteries include the systemic (sis-TEM-ik) arteries that supply blood to all regions of the body and the pulmonary (PUL-mo-ner-e) arteries that supply blood to the lungs. Blood leaving the heart through the aorta and returning to the right atrium is said to be in the systemic loop. Blood leaving the heart via the pulmonary trunk and returning to the left atrium is in the pulmonary loop. Note that blood in the pulmonary arteries is considered arterial blood even though it is deoxygenated. This is because the flow is away from the heart.

Branches of the Aorta

After the names of the arteries listed below, *describe* the area of the body supplied by each vessel and *label* Figure 12.7.

aorta

  brachiocephalic artery

    right subclavian (sub-KLA-ve-an) artery

      right vertebral artery

      right internal thoracic (tho-RAS-ik) artery

      right anterior intercostal (in-ter-KOS-tal) arteries

    *right common carotid (kah-ROT-id) artery

  *left common carotid artery

  left subclavian artery

    left vertebral (VER-te-bral) artery

  posterior intercostal (in-ter-KOS-tal) artery

  celiac (SE-le-ak) trunk

  superior mesenteric (mes-en-TER-ik) artery

  renal (RE-nal) arteries

  gonadal (go-NAD-al [ovarian, testicular]) arteries

  inferior mesenteric artery

  right common iliac (IL-e-ak) artery

    right internal iliac artery

    right external iliac artery

  left common iliac artery

    left internal iliac artery

    left external iliac artery

    left internal thoracic artery

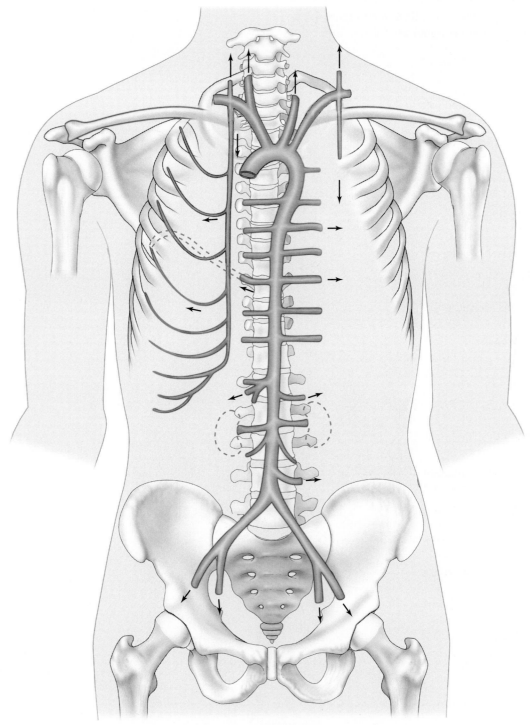

FIG. 12.7 Branches of the aorta, anterior view.

Arteries of the Head

After the arteries listed below, *describe* the area of the body supplied by each vessel, then *label* Figure 12.8. The printed labels on the figure are for reference.

aorta

    right common carotid artery

        right internal carotid artery

        right external carotid artery

    right subclavian artery

        right vertebral artery

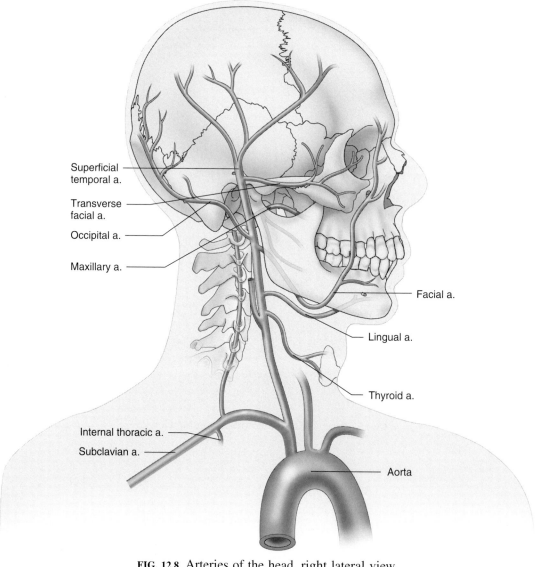

FIG. 12.8 Arteries of the head, right lateral view.

Arteries of the Brain

Figure 12.9 illustrates the major arteries of the ventral surface of the brain. This diagram is primarily for reference but you should *locate* the following:

internal carotid arteries

vertebral arteries

cerebral arterial circle

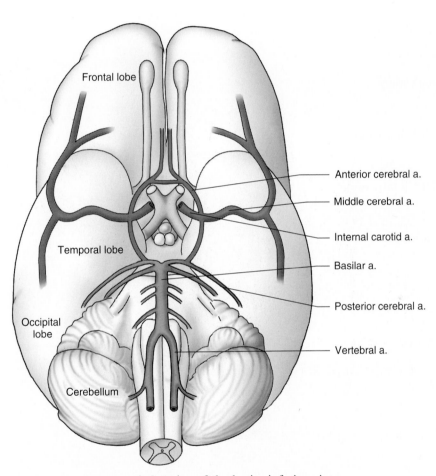

FIG. 12.9 Arteries of the brain, inferior view.

Arteries of the Superior Appendage

After each of the arteries listed below, *describe* the regions of the superior appendage that are supplied by the vessel and *label* Figure 12.10.

subclavian artery

*axillary (AK-sih-l<u>ar</u>-<u>e</u>) artery

brachial (BRA-k<u>e</u>-al) artery

deep brachial artery

ulnar (UL-nar) artery

palmar (PAL-mar) arch

digital (DIJ-ih-tal) arteries

*radial (R<u>A</u>-d<u>e</u>-al) artery

palmar arch

digital arteries

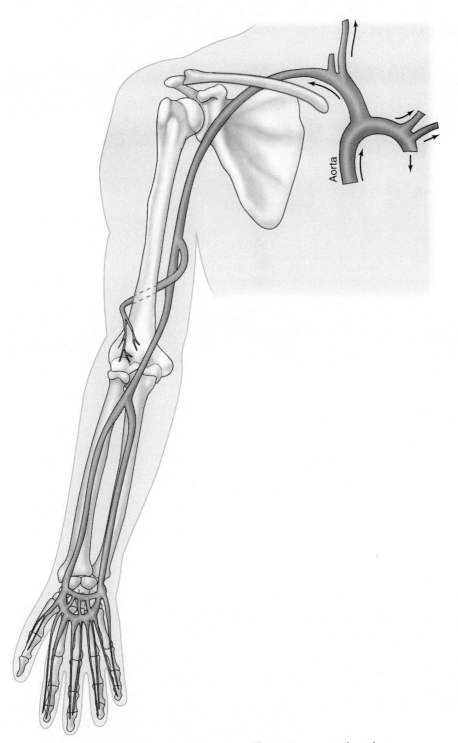

Aorta

FIG. 12.10 Arteries of the superior appendage, anterior view.

Arteries of the Inferior Appendage

After each of the arteries listed below, *describe* the regions of the inferior appendage that are supplied by the vessel and *label* Figure 12.11.

external iliac artery

femoral (FEM-or-al) artery

deep femoral artery

popliteal (pop-LIT-e-al) artery

anterior tibial (TIB-e-al) artery

dorsalis pedis (dor-SA-lis PE-dis) artery

deep plantar (PLAN-tahr) artery

digital arteries

posterior tibial artery

medial and lateral plantar arteries

digital arteries

peroneal (per-O-ne-al) artery

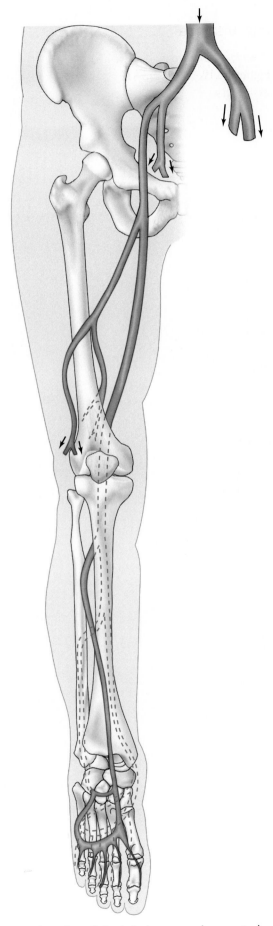

FIG. 12.11 Arteries of the inferior appendage, anterior view.

Arteries of the Abdominal Viscera

After each of the arteries below, *describe* the regions of the viscera supplied. *Refer* to Figure 12.12. This diagram is for reference only unless otherwise specified by your instructor.

celiac trunk

splenic (SPLEN-ik) artery

hepatic (he-PAT-ik) artery

gastric (GAS-trik) artery

gastroepiploic (gas-tro-ep-ih-PLO-ik)

inferior pancreaticoduodenal
(pan-kre-at-ih-ko-doo-AH-deh-nal) artery

superior mesenteric (mes-en-TER-ik) artery

middle colic (KOL-ik) artery

right colic artery

ileocolic (il-e-o-KOL-ik) artery

jejunal (je-JOO-nal) artery

ileal (IL-e-al) artery

inferior mesenteric artery

inferior left colic artery

superior rectal (REK-tal) artery

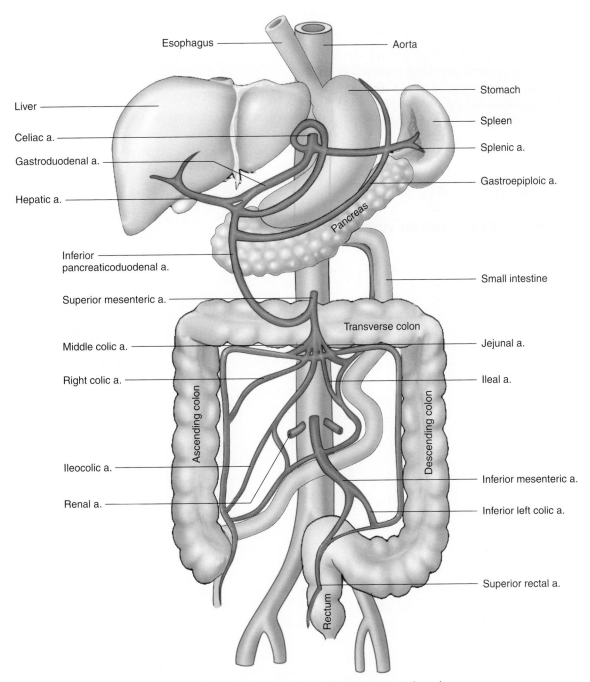

Esophagus

Aorta

Liver

Stomach

Celiac a.

Spleen

Gastroduodenal a.

Splenic a.

Hepatic a.

Gastroepiploic a.

Pancreas

Inferior
pancreaticoduodenal a.

Small intestine

Superior mesenteric a.

Transverse colon

Middle colic a.

Jejunal a.

Right colic a.

Ileal a.

Ascending colon

Descending colon

Ileocolic a.

Inferior mesenteric a.

Renal a.

Inferior left colic a.

Superior rectal a.

Rectum

FIG. 12.12 Arteries of the abdominal viscera, anterior view.

THE VEINS

Veins are vessels that carry blood **toward** the heart. These include the systemic veins that drain blood from all regions and organs of the body and the pulmonary veins that drain blood from the lungs. Note that blood returning to the left atrium from the pulmonary loop is considered to be venous blood even though it is high in oxygen. This is because the flow is toward the heart. Backflow, due to the effects of gravity, is prevented by valves within the vessels below the neck.

Tributaries to the Superior and Inferior Vena Cava

After each vein listed below, *describe* the area of the body drained and *label* Figure 12.13.

superior vena cava

    right and left brachiocephalic veins

    azygos (AZ-ih-gos) vein

      posterior intercostal veins

inferior vena cava

    hepatic vein

    right renal vein

    right gonadal vein

    left renal vein

      left gonadal vein

    common iliac veins

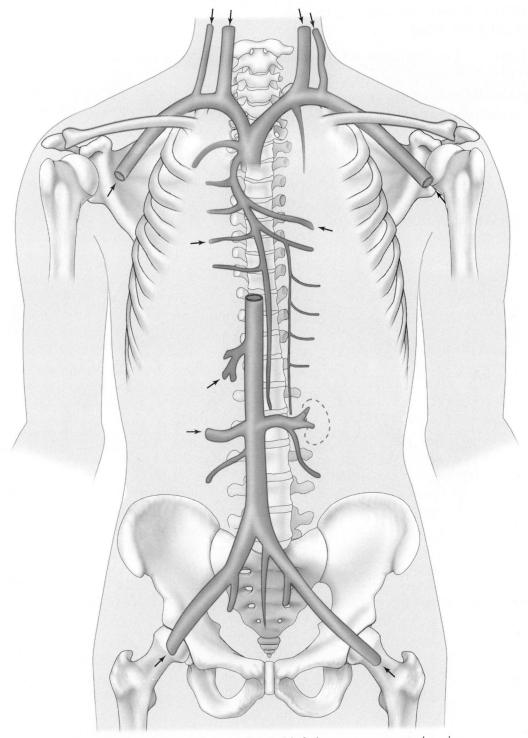

FIG. 12.13 Tributaries to the superior and inferior vena cava, anterior view.

Veins of the Brain, Tributaries to the Internal Jugular Vein

Most of Figure 12.14 is for reference but you should *locate* and *label* the following:

internal jugular (JUG-u-lar) vein

    sigmoid sinus (SIG-moyd SI-nus)

        transverse sinus

            superior sagittal (SAJ-ih-tal) sinus

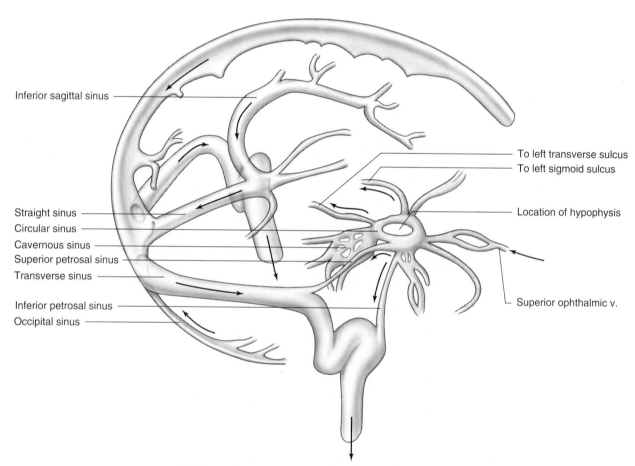

FIG. 12.14 Tributaries to the internal jugular veins, right lateral view.

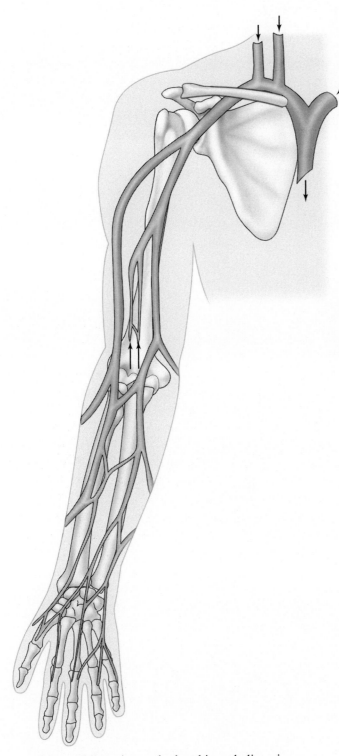

FIG. 12.15 Tributaries to the brachiocephalic veins, anterior view.

Veins of the Superior Appendage

Tributaries to the Brachiocephalic Veins

After each of the veins listed below, *describe* the area of the body drained and *label* Figure 12.15.

left and right brachiocephalic vein

external jugular vein

internal jugular vein

subclavian vein

axillary vein

*cephalic (seh-FAL-ik) vein

*basilic (bah-SIL-ik) vein

*median cubital (KYOO-bih-tl) vein (used for venipuncture)

brachial vein

radial vein

ulnar vein

Veins of the Inferior Appendage

Tributaries to the Common Iliac Veins

After each of the veins listed below *describe* the area of the body drained and *label* Figure 12.16.

common iliac veins

internal iliac (hypogastric) vein

external iliac vein

femoral vein

deep femoral vein

great saphenous (sah-FE-nus) vein

dorsal venous arch

digital veins

popliteal vein

small saphenous vein

dorsal venous arch

digital veins

anterior tibial vein

dorsal venous arch

digital veins

posterior tibial vein

peroneal vein

plantar arches

digital veins

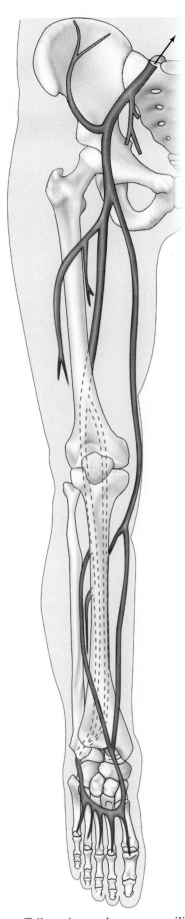

FIG. 12.16 Tributaries to the common iliac vein, anterior view.

HEPATIC PORTAL CIRCULATION

After each of the vessels listed below, *describe* the area of the body drained and *label* Figure 12.17.

inferior vena cava

hepatic vein

liver sinusoids (SI-nyoo-soyds)

hepatic portal (POR-tl) vein

gastric veins

splenic vein

inferior mesenteric vein

superior mesenteric vein

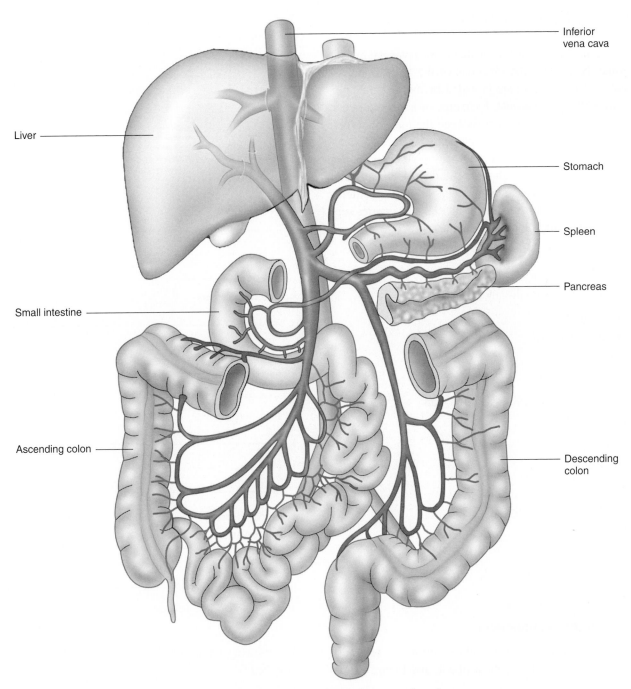

Inferior vena cava

Liver

Stomach

Spleen

Pancreas

Small intestine

Ascending colon

Descending colon

FIG. 12.17 Hepatic portal circulation, anterior view.

FETAL CIRCULATION

Fetal circulation differs from that in the postnatal human primarily because the functions of the lungs, kidneys, and digestive system are provided by the maternal systems across the placenta. Exchange of gases, nutrients, and waste products occurs between the maternal blood in the endometrium and fetal blood in the placenta. Oxygenated, nutrient-rich blood flows through the umbilical vein to the liver, then through the ductus venosus into the inferior vena cava and on to the heart. Two umbilical arteries branching from the internal iliac arteries supply blood returning to the placenta.

Locate and *label* the following in Figure 12.18.

placenta (plah-SEN-tah)

umbilical (um-BIL-ih-kal) arteries

umbilical vein

inferior vena cava

aorta

liver

ductus venosus (DUK-tus veh-NO-sus)

right atrium

right ventricle

left atrium

left ventricle

foramen ovale (O-val-e)

pulmonary artery

lungs

ductus arteriosus (ar-te-re-O-sus)

superior vena cava

CRITICAL THINKING

1. *Describe* the route that blood follows in fetal circulation. *Consult* your textbook and Figure 12.18.

_____

_____

_____

_____

_____

_____

_____

2. *Describe* the fate of the following fetal structures in the adult.

umbilical arteries

_____

_____

_____

umbilical vein

_____

_____

_____

ductus venosus

_____

_____

_____

foramen ovale

_____

_____

_____

_____

ductus arteriosus

_____

_____

_____

3. Following labor and delivery, what is the placenta called?

_____

_____

_____

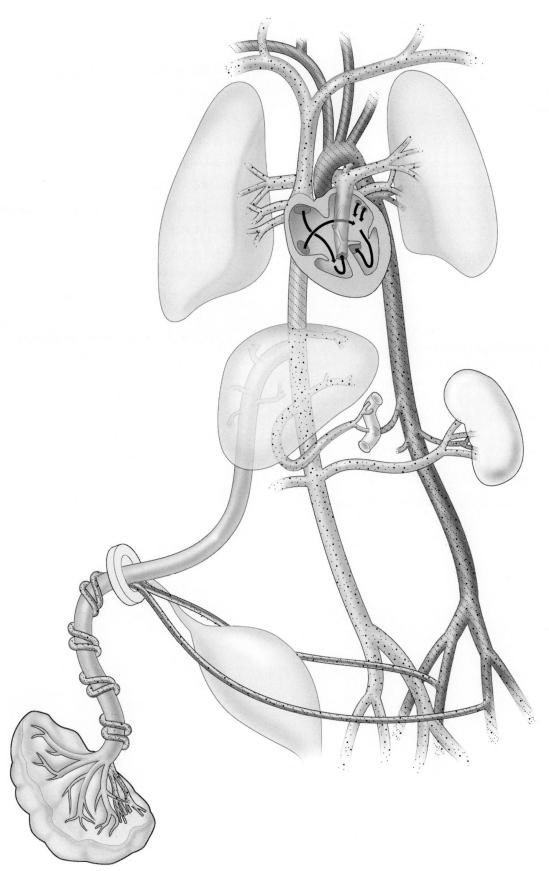

FIG. 12.18 Fetal circulation schematic, anterior view.

4. Which vessel of fetal circulation carries the most oxygen?

_____

5. *Define:*

hypertension

_____

_____

_____

aneurysm

_____

_____

_____

Test your knowledge of the circulatory system by *completing* the following problems on a separate sheet of paper. *Assume* the adult pattern of circulation. *List* all of the vessels, chambers, and valves in the order that they are encountered. Use a separate sheet of paper for your answers.

6. Begin: aorta

End: superior vena cava

Pass through: digital vein of the right hand

7. Begin: aorta

End: superior vena cava

Pass through: superior sagittal sinus

8. Begin: aorta

End: inferior vena cava

Pass through: left ovarian capillaries

9. Begin: aorta

End: aorta

Pass through: capillaries of the peroneal artery

10. Begin: right coronary artery

End: inferior vena cava

Pass through first: descending colon capillaries

Pass through second: right renal capillaries

_____ **11.** At what level does gas exchange with the tissues occur?

a. capillaries   c. arteries
b. veins   d. arterioles

_____ **12.** Which best describes the arteries?

a. carry blood toward the heart
b. contain valves
c. pressures rarely rise above 2mmHg
d. all branch from the vena cava

_____ **13.** The vessels with the smallest lumen relative to the thickness of the wall are

a. veins.   c. capillaries.
b. venules.   d. arteries.

Define the Greek or Latin root for the following terms:

14. ventricle

_____

_____

_____

15. atrium

_____

_____

_____

16. aorta

_____

_____

_____

17. cava

_____

_____

_____

18. brachiocephalic

_____

_____

_____

19. renal

_____

_____

_____

20. jugular

_____

_____

_____

_____

BLOOD

Blood is a connective tissue and consists of a liquid matrix, the plasma, and cellular blood solids. Blood functions to transport gases, nutrients, and hormones from one part of the body to another. At the same time, blood also carries heat, ions, and buffers to regulate body temperature and blood pH. Finally, blood carries special cells and antibodies that are important in the control of disease.

Blood solids (formed elements) consist of the following:

Erythrocytes (eh-RITH-ro-sits)—red blood cells

Leukocytes (LOO-ko-sits)—white blood cells

Agranulocytes (a-GRAN-yoo-lo-sits)—contain no cytoplasmic granules

Lymphocytes (LIM-fo-sits)—produce antibodies and have a large, round nucleus

Monocytes (MON-o-sits)—phagocytic cells with a large kidney-bean-shaped nucleus

Granulocytes (GRAN-yoo-lo-sits)—have cytoplasmic granules and lobed nuclei

Basophils (BA-so-fils)—involved in allergic reactions, having bilobed nuclei and staining blue with basic dyes

Eosinophils (e-o-SIN-o-fils)—attack antigen-antibody complex, having bilobed nuclei and staining red with acid dyes

Neutrophils (NOO-tro-fils)—phagocytic cell with multilobed nuclei

Platelets (PLAT-lets)—tiny blood solids, about 3μ in diameter, that are important in blood clotting

Erythrocytes, granulocytes, and platelets are produced in the red bone marrow. Agranulocytes also originate in the bone marrow but the monocytes mature in the peripheral tissues and many lymphocytes migrate to the thymus and eventually to the lymph nodes. *Observe* a prepared slide of human blood. *Locate,* with the help of Figure 12.19 and your textbook, the blood solids listed above.

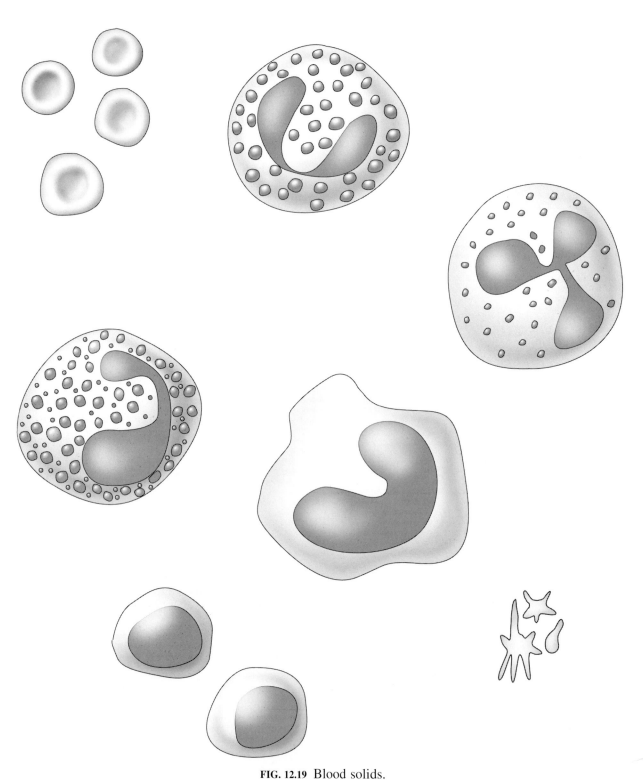

FIG. 12.19 Blood solids.

SHORT ANSWER

1. What is hemoglobin?

2. What is hematocrit?

3. *List* five major components of blood plasma.

4. *List* each blood solid and provide a function.

5. Which is the most abundant blood solid? The least abundant?

MATCHING

Match the descriptions in column A with the terms in column B.

A

_____ 1. heart muscle

_____ 2. outer membrane surrounding the heart

_____ 3. valve prevents backflow to ventricle

_____ 4. valve between left atrium and ventricle

_____ 5. pumps blood to the systemic loop

_____ 6. transfers contractile impulse to ventricle

_____ 7. vein on left side of heart muscle

_____ 8. vein on right side of heart muscle

_____ 9. initiates heart contraction

_____ 10. chamber which pumps blood to lungs

_____ 11. systemic vein carrying blood to heart

_____ 12. supplies blood to heart muscle

_____ 13. collects blood from veins of heart

B

a. atrioventricular node

b. atrioventricular valve (left)

c. coronary artery

d. coronary sinus

e. fibrous pericardium

f. great cardiac vein

g. left ventricle

h. myocardium

i. sinoatrial node

j. semilunar valve

k. small cardiac vein

l. right ventricle

m. vena cava

Match the descriptions in column A with the terms in column B.

A

_____ 1. lymphocytes and monocytes

_____ 2. water plus solutes

_____ 3. blood production

_____ 4. neutrophils, eosinophils, and basophils

_____ 5. general term for all white blood cells

_____ 6. agranulocyte involved in antibody immunity

_____ 7. tiny, anuclear solid, involved in clotting

_____ 8. oxygen-transporting blood solid

B

a. agranulocyte

b. granulocyte

c. erythrocyte

d. hemopoiesis

e. leukocyte

f. lymphocyte

g. plasma

h. thrombocyte

Match the vessels in column A with the largest area supplied or drained in column B.

A

__l__ 1. celiac trunk

__j__ 2. gonadal artery

__h__ 3. femoral artery

__e__ 4. common carotid artery

__b__ 5. internal jugular vein

__c__ 6. inferior mesenteric artery

__d__ 7. radial artery

__g__ 8. renal artery

__k__ 9. cephalic vein

__a__ 10. hepatic portal vein

__h__ 11. posterior tibial artery

__i__ 12. pulmonary artery

B

a. all intestines

b. brain

c. colon

d. forearm

e. head

f. inferior appendage

g. kidney

h. leg

i. lung

j. ovary

k. superior appendage

l. superior intestines

MULTIPLE CHOICE

_____ 1. Which of the following connects the systemic and pulmonary vessels in the fetus?

    a. foramen ovale
    b. ductus venosus
    c. umbilical vein
    d. ductus arteriosus
    e. none of the above

_____ 2. Which of the following connects the atria in the fetal heart?

    a. foramen ovale
    b. ductus venosus
    c. umbilical artery
    d. ductus arteriosus
    e. more than one of the above

_____ 3. Of the vessels below, which carries blood of the highest oxygen content in the fetus?

    a. umbilical vein
    b. umbilical artery
    c. thoracic aorta
    d. internal iliac vein
    e. pulmonary vein

_____ 4. Venous blood flows from intestines to liver via the

    a. renal vein.
    b. hepatic vein.
    c. inferior mesenteric artery.
    d. celiac trunk.
    e. none of the above.

_____ 5. The pacemaker of the heart is/are the

    a. conduction myofibers.
    b. atrioventricular node.
    c. atrioventricular bundle.
    d. sinoatrial node.
    e. none of the above.

_____ 6. Which of the following drains to the right atrium?

    a. coronary sinus
    b. superior vena cava
    c. inferior vena cava
    d. pulmonary vein
    e. more than one of the above

_____ 7. How do veins differ from arteries of the same external diameter?

  a. veins lack a tunica interna
  b. veins have a smaller lumen
  c. veins never have muscle
  d. veins may have valves
  e. more than one of the above

_____ 8. Which of the following best describes a capillary?

  a. tunica media only
  b. tunica externa only
  c. tunica interna only
  d. muscular wall
  e. more than one of the above

_____ 9. Which of the following does NOT have three cusps?

  a. pulmonary semilunar valve
  b. right atrioventricular valve
  c. aortic semilunar valve
  d. left atrioventricular valve

_____ 10. Which one of the following is not a formed element in blood?

  a. granulocyte   c. globulin
  b. platelet   d. erythrocyte

_____ 11. Which one of the following is a function of blood?

  a. transport waste products from cells of the body to the kidneys
  b. transport food material
  c. regulate extracellular environment
  d. all are functions of blood

_____ 12. The route of blood from the right side of the heart to the lungs and back to the left side of the heart is called the _____ loop.

  a. pulmonary   c. lymphatic
  b. cardiac   d. systemic

Cardiovascular System of the Cat

13

Contents

13

Objectives

1. Dissect the cat cardiovascular system to better understand that of humans.

2. Be able to list the differences between the cardiovascular pattern of the cat and that of the human.

OPENING THE BODY CAVITY

Begin the dissection of the cat cardiovascular system by *making* a midventral incision through the body wall. *Start* at the anterior end of the sternum and *continue* to the pubis. Cutting through the sternum may be difficult; *use* a stout scissors. Then *make* two lateral incisions through the diaphragm by *cutting* close to the body wall on each side. You may find that *pinning* or *tying* the body cavity open is helpful.

Note: The arteries have been filled with red latex, the veins with blue latex. If you have a "triple-injected" animal, the hepatic portal system will contain yellow latex. As you *search* for the vessels within connective tissue, be careful not to damage them.

It is best to *begin* by tracing vessels (both arteries and veins) away from the heart. Doing so provides a landmark where you can begin. Since veins have thinner walls and are more easily destroyed, *look* for them first.

To approach the superior vena cava and the aorta, it is necessary to carefully *scrape* away the thymus gland. It lies superior to the heart on the ventral side. Once you locate the aorta you will need to *pull* the left lung to the right so that you can see the vessel as it passes down the dorsal (posterior) wall of the body cavity. *Cut* along the diaphragm to the point where the aorta passes through it. Now, *deflect* the viscera to the right so that you can clearly see the aorta and inferior vena cava.

Note: The inferior vena cava passes dorsal to the liver and is difficult to follow until you locate it caudal to the liver. However, you should *trace* this vessel from the heart to the cranial end of the liver.

The hepatic portal system is often injected with yellow latex. As you *trace* these yellow vessels among the intestines, they will finally join to form a single yellow vessel entering the liver. Sinusoids and hepatic veins are buried within the liver and you will need to *cut* across one of the lobes to see them.

Vessels in this chapter are listed below in outline form. Therefore, where arteries are listed, indentations indicate branches. That is, the brachiocephalic and left subclavian arteries are branches of the aorta. Similarly the right subclavian, right common carotid, and left common carotid arteries are branches of the brachiocephalic artery (see immediately below). When veins are listed, a vessel indented below another indicates that the indented vessel is a tributary to the one listed above.

ARTERIES CRANIAL TO THE HEART

Locate the following vessels on the cat. Refer to Figure 13.1 and CDA-12.

aorta

    brachiocephalic artery

        right subclavian artery

            right vertebral artery

            axillary artery

                brachial artery

                    ulnar artery

                    radial artery

        right common carotid artery

            right internal carotid artery

            right external carotid artery

        left common carotid artery (a branch of the aorta in humans)

            left internal carotid artery

            left external carotid artery

    left subclavian artery

        left vertebral artery

        left internal thoracic artery

            left anterior intercostal arteries

VEINS CRANIAL TO THE HEART

Locate the following vessels on the cat. Refer to Figure 13.1 and CDA-12.

superior vena cava

    azygos vein

        posterior intercostal veins

    right and left brachiocephalic veins

        left internal thoracic vein

            anterior intercostal veins

        external jugular veins

        internal jugular veins

        subclavian veins

            axillary veins

    right internal thoracic vein

        anterior intercostal veins

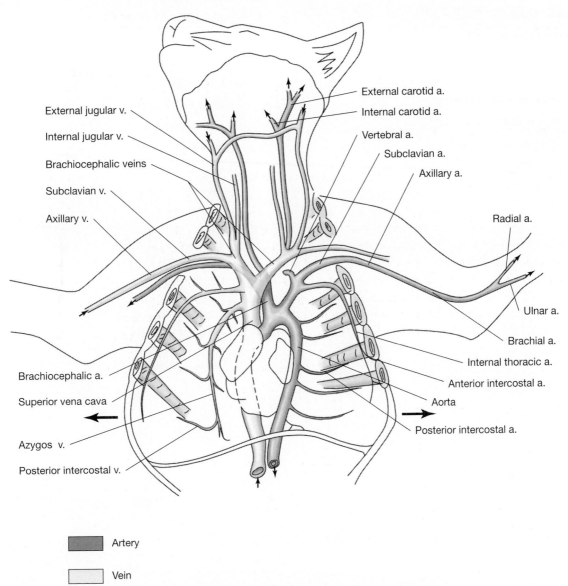

External jugular v.

Internal jugular v.

Brachiocephalic veins

Subclavian v.

Axillary v.

External carotid a.

Internal carotid a.

Vertebral a.

Subclavian a.

Axillary a.

Radial a.

Ulnar a.

Brachial a.

Internal thoracic a.

Anterior intercostal a.

Aorta

Posterior intercostal a.

Brachiocephalic a.

Superior vena cava

Azygos v.

Posterior intercostal v.

Artery

Vein

FIG. 13.1 Cat, arteries and veins cranial to the heart, ventral view.

ARTERIES CAUDAL TO THE HEART

Locate the following vessels on the cat. Refer to Figure 13.2 and CDA-13.

aorta

    celiac trunk

    superior mesenteric artery

    renal arteries

    gonadal (ovarian, testicular) arteries

    inferior mesenteric artery

    common iliac artery

      internal iliac arteries

      external iliac arteries

        femoral artery

VEINS CAUDAL TO THE HEART

Locate the following vessels on the cat. Refer to Figure 13.2 and CDA-13.

inferior vena cava

    hepatic vein

    right renal vein

    right gonadal vein

    left renal vein

      left gonadal vein

    common iliac veins

      internal iliac veins

      external iliac veins

        femoral vein

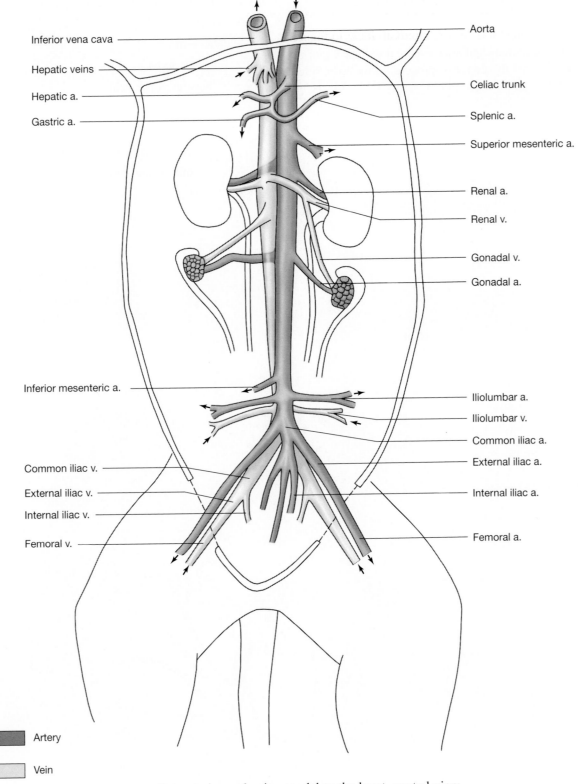

Inferior vena cava

Hepatic veins

Hepatic a.

Gastric a.

Aorta

Celiac trunk

Splenic a.

Superior mesenteric a.

Renal a.

Renal v.

Gonadal v.

Gonadal a.

Inferior mesenteric a.

Iliolumbar a.

Iliolumbar v.

Common iliac a.

External iliac a.

Common iliac v.

External iliac v.

Internal iliac v.

Femoral v.

Internal iliac a.

Femoral a.

Artery

Vein

FIG. 13.2 Cat, arteries and veins caudal to the heart, ventral view.

HEPATIC PORTAL CIRCULATION

The **hepatic portal** (POR-tal) **circulation** carries blood from the intestinal capillaries, where it will have picked up products of the digestive process, to the liver sinusoids (SI-nyoo-soyds).

Locate the following vessels in Figure 13.3.

inferior vena cava

    hepatic vein

        liver sinusoids

            hepatic portal vein

                gastric vein

                splenic vein

                superior mesenteric vein

                inferior mesenteric vein

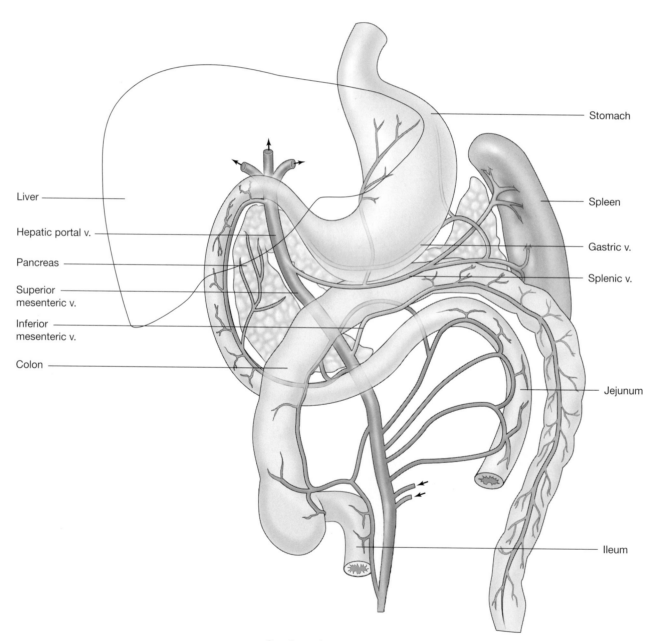

FIG. 13.3 Cat, hepatic portal circulation, ventral view.

14 Lymphatic System

Contents

Objectives

1. Describe the components of the lymphatic system and provide their functions.

2. Describe the flow pattern of the lymphatics beginning at a lymph capillary and terminating in a lymphatic duct.

3. Describe the structure of a lymph node and provide a function for each part.

4. List the primary lymph nodes.

5. List the lymph organs and describe their locations and functions.

FUNCTIONS OF THE LYMPHATIC SYSTEM

The lymphatic system includes the fluid called lymph (limf) as well as the lymph vessels, tonsils (TON-sils), spleen (splen), and thymus (THI-mus). The main function of this system is to return fluids, which have left the blood capillaries and entered the interstitial spaces, to the bloodstream. The lymphatic system is also responsible for absorption and transport of fats from the digestive system to the bloodstream. Finally, the lymphatic system functions to fight disease.

LYMPH VESSELS

Locate and *label* the following in Figure 14.1.

left lymphatic (thoracic) duct

    cisterna chyli (sis-TER-nah KĪ-lī)

right lymphatic duct

lymph nodes

    cervical

    axillary

        mammary plexus (MAM-er-ē PLEK-sus)

    intestinal

    inguinal (ING-gwih-nal)

    popliteal

lymph vessels

    lacteals (LAK-tē-lz)

afferent (AF-er-ent) lymphatics

efferent (EF-er-ent) lymphatics

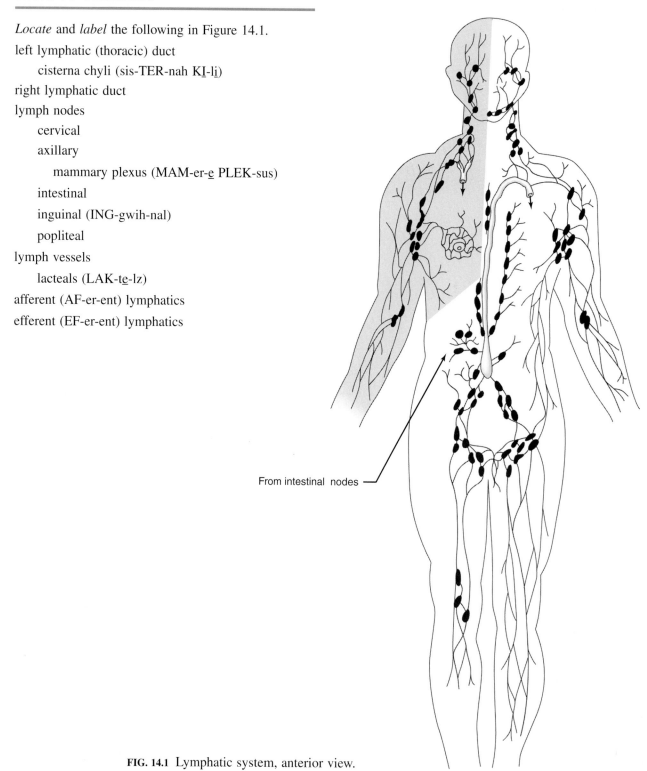

From intestinal nodes

FIG. 14.1 Lymphatic system, anterior view.

LYMPH NODE

Locate and *label* the following in Figure 14.2.

afferent lymphatic vessel

efferent lymphatic vessel

valve

cortex (KOR-teks)

    germinal (JER-mih-nal) center

    lymphatic nodule

    cortical (KOR-tih-kl) sinus

medulla (meh-DUL-ah)

    medullary cord

    medullary sinus

capsule

trabeculae

LYMPH ORGANS

As you refer to a diagram of the lymphatic system in your textbook, *locate* the following organs:

spleen

thymus

tonsils

    pharyngeal (adenoid)

    palatine

    inguinal

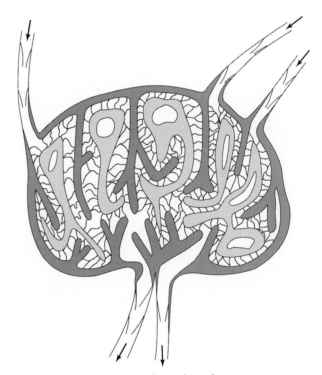

FIG. 14.2 Lymph node.

DISSECTION OF THE CAT LYMPHATICS

Locate the following lymphatic tissues of the cat.

lymph nodes

    axillary

    cervical

    inguinal

    intestinal

spleen

thymus

SHORT ANSWER

1. *Define* the following terms:

lymphangitis

metastasis

edema

T cell

2. *Trace* a drop of lymph from the interstitial spaces of the foot to the left subclavian vein. *List* all of the lymph vessels and organs in the order of occurrence.

MATCHING

Match the descriptions in column A with the terms in column B.

A

_____ 1. collects lymph from tissues

_____ 2. main lymph vessel

_____ 3. an intestinal lymph capillary

_____ 4. agranulocyte produced here

_____ 5. a result of lymphatic insufficiency

B

a. edema

b. lacteal

c. lymph capillary

d. lymph node

e. thoracic duct

MULTIPLE CHOICE

_____ 1. The lymphatic vessels leading away from a lymph node are called

    a. afferent lymphatic vessels.
    b. lymph trunks.
    c. lymphatic ducts.
    d. efferent lymphatic vessels.

_____ 2. Which one of the following is NOT true of the lymphatic system?

    a. provides defense against invasion of foreign particles
    b. returns interstitial material back to the bloodstream
    c. empties into the arterial system
    d. forms leukocytes

_____ 3. The tonsils located in the posterior wall of the nasopharynx are the

    a. palatine tonsils.
    b. pharyngeal tonsils.
    c. lingual tonsils.
    d. submandibular tonsils.

15 Nervous System

Contents

15

CHAPTER

Objectives

1. Describe the anatomy of the neuron.

2. Define and describe polarization, depolarization, and repolarization of the neuron.

3. Describe the anatomy and function of the synapse.

4. Understand the nature and function of neurotransmitters.

5. List and describe the function of classes of neuronal pathways.

6. Describe the function and location of interneurons.

7. Describe the simple spinal reflex arc.

8. Define the terms PNS, CNS, and ANS.

9. Describe the organization of the cerebrum, cerebellum, and spinal cord with reference to gray and white matter.

10. List the divisions and subdivisions of the brain.

11. Describe the locations and functions of the primary motor, primary sensory, and association areas of the cerebrum.

12. Name and number each of the cranial nerves. Describe the path by which each leaves the cranium. State if each nerve is sensory or mixed.

13. Name the branches of the trigeminal nerve. Describe the areas each innervates. Trace the path by which each branch leaves the cranium.

14. Describe the anatomy of the spinal cord. Differentiate the locations and functions of the roots. Describe a plexus.

15. Describe the anatomical and functional differences of the two divisions of the autonomic nervous system. Where does each originate?

16. Dissect the sheep brain to better understand that of humans.

DIVISIONS OF THE NERVOUS SYSTEM

The nervous system is composed of specialized cells called **neurons** (NOO-rons). A group of parallel neurons is a **nerve**. The function of the neuron is to conduct information (depolarization and repolarization events) from one point to another within the body.

The nervous system is divided into two major anatomical regions. The **peripheral nervous system (PNS)** is made up of peripheral nerves and receptors. These include the cranial and spinal nerves. The **central nervous system (CNS)** is the brain and spinal cord.

FUNCTIONAL CLASSES OF NEURONS

There are six functional classes of neurons in the nervous system:

1. **somatic** (so-MAT-ik) **sensory**
2. **somatic motor**
3. **visceral** (VIS-er-al) **sensory**
4. **visceral motor** (2 classes)
   a. **parasympathic** (par-ah-sim-pah-THET-ik)
   b. **sympathetic** (sim-pah-THET-ik)
5. **association** or **interneurons**

Motor neurons are also referred to as **efferent** (EF-er-ent) or **descending neurons** and the sensory neurons as **afferent** (AF-er-ent) or **ascending neurons**. The two types of visceral efferents are called the **autonomic** (aw-to-NOM-ik) **nervous system** or **ANS**. These two visceral motor divisions are often antagonistic.

HISTOLOGY OF A MYELINATED, MULTIPOLAR NEURON

Locate and *label* the following features in Figure 15.1. For each term listed below, *provide* a brief description of the function.

cell body _____

_____

dendrite (DEN-drit) _____

_____

axon (AX-on) _____

_____

nucleus _____

_____

neurolemmocytes (noo-ro-LEH-mo-sit) ___

_____

node of Ranvier (RON-ve-a) _____

_____

axon terminals _____

_____

synaptic (sih-NAP-tik) bulbs _____

_____

synaptic vesicles _____

_____

CRITICAL THINKING

1. What substance is responsible for the white appearance of nerve axons and the white matter in the brain and spinal cord?

   _____

2. What are the three structural classes of neurons? What are their anatomical locations in the nervous system?

   _____

   _____

   _____

3. *List* the six functional classes of neurons and indicate the general function of each.

   _____

   _____

   _____

   _____

   _____

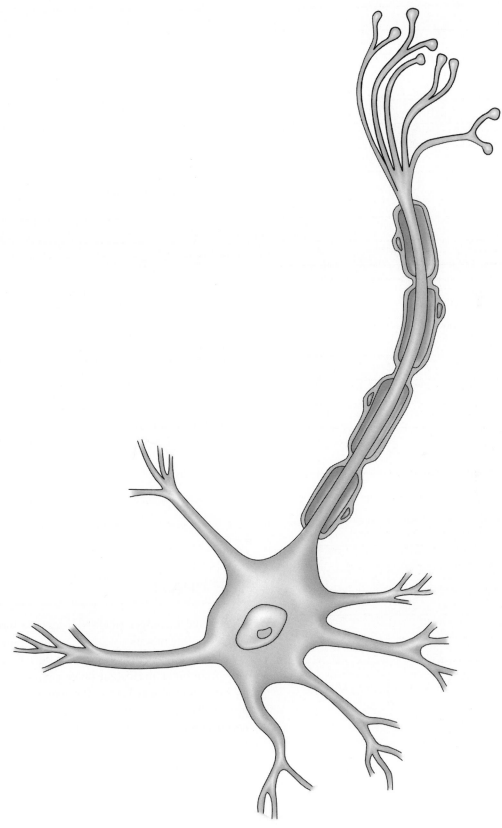

FIG. 15.1 Myelinated, multipolar neuron.

4. *Define* the following terms:

a. nerve impulse

_____

_____

_____

_____

b. stimulus

_____

_____

_____

c. excitability

_____

_____

_____

_____

d. synapse

  (i) axodendritic

_____

_____

_____

_____

  (ii) axosomatic

_____

_____

_____

_____

  (iii) axoaxonic

_____

_____

_____

_____

e. synaptic cleft

_____

_____

_____

_____

f. neurotransmitter substance

  (i) acetylcholine

_____

_____

_____

  (ii) norepinephrine

_____

_____

_____

g. presynaptic/postsynaptic neuron

_____

_____

_____

h. myoneural junction

_____

_____

_____

THE BRAIN

The cranial meninges (meh-NIN-jez)—a three-layered complex of membranes—surrounds the brain. The outer layer, the dura mater (DOO-rah MA-ter), is actually two layers: an outer **periosteal layer** against the skull and an inner **meningeal layer**. The meninges support and protect the delicate nervous tissues of the brain. The **cerebrospinal** (SER-re-bro-SPI-nal) **fluid** (**CSF**) circulates through the subarachnoid space located between the arachnoid and pia (PI-a) mater, the deepest of the meninges. The CSF absorbs shock, provides some nutrients, and removes waste from the CNS.

THE CRANIAL MENINGES

Locate and *label* the following in Figure 15.2.

meninges (singular menix [MEH-ninks])

 dura mater

  falx cerebri (falks SER-eh-br<u>e</u>)

 arachnoid membrane

  arachnoid villi (a-RAK-noyd VIL-<u>i</u>)

  subarachnoid space

 pia mater

other structures

 left and right cerebral hemisphere

 sagittal sinus

 scalp

 bone

BRAIN REGIONS

In the adult, the brain is divided into the following four major regions and their subdivisions:

1. brain stem

  medulla oblongata (meh-DUL-ah ob-long-GAH-tah)

  pons (ponz)

  mesencephalon (mez-en-SEF-ah-lon)

2. diencephalon (d<u>i</u>-en-SEF-ah-lon)

  thalamus (THAL-ah-mus)

  hypothalamus

3. cerebrum

4. cerebellum (ser-eh-BEL-um)

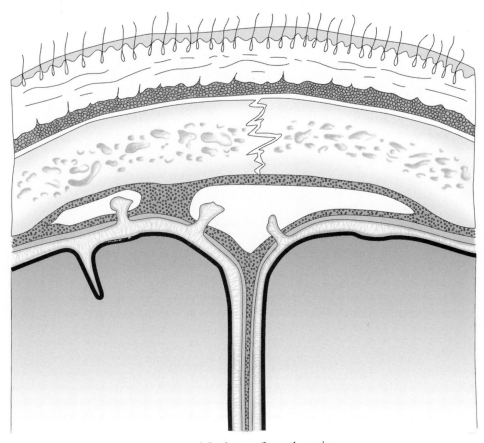

FIG. 15.2 Meninges, frontal section.

Brain, Right Lateral View

Locate and *label* the following in Figure 15.3.

spinal cord

brain stem

    medulla oblongata

    pons

cerebellum

cerebrum

    central sulcus (SUL-kus)

    lateral sulcus

    postcentral gyrus (JI-rus)

    precentral gyrus

    transverse fissure

    frontal lobe

    occipital lobe

    temporal lobe

    parietal lobe

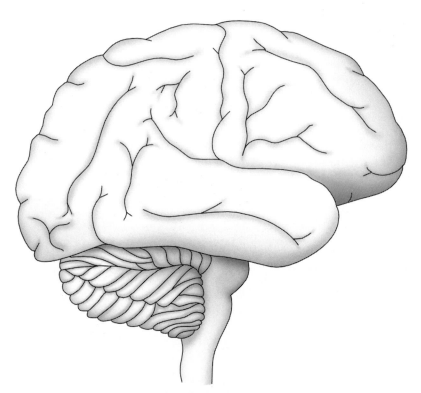

FIG. 15.3 Brain, right lateral view.

Brain, Midsagittal View, Left Hemisphere

Locate and *label* the following in Figure 15.4.

lateral ventricle

third ventricle

    cerebral aqueduct (AK-we-dukt)

    choroid plexus (KO-royd)

fourth ventricle

spinal cord

    central canal

brain stem

    medulla oblongata

    pons

    mesencephalon (midbrain)

       tectum (TEK-tum)

       corpora quadrigemina (KOR-po-rah qwod-rih-JEM-ih-nah)

         superior and inferior colliculi (ko-LIK-u-li)

cerebellum

    arbor vitae (white matter) (AR-bor VI-te)

    cerebellar nuclei (gray matter)

    folia (FO-le-ah)

diencephalon

    thalamus

      intermediate mass

    hypothalamus

      infundibulum

    hypophysis (pituitary gland) (hi-POF-ih-sis)

    pineal (PIN-e-al) gland

cerebrum

    cortex (KOR-teks)

    medulla

    corpus callosum (KOR-pus kah-LO-sum)

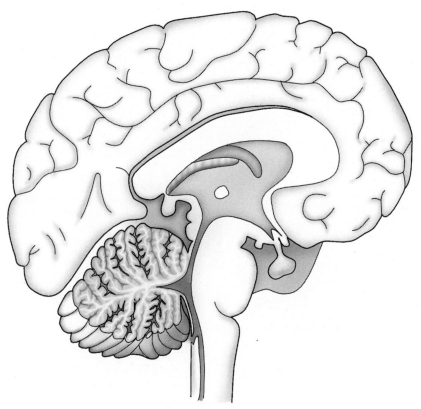

FIG. 15.4 Brain, midsagittal view, left hemisphere.

Brain, Frontal Section Through the Hypophysis

Locate and *label* the following in Figure 15.5.

diencephalon
   hypothalamus
      infundibulum
      hypophysis
   thalamus
      intermediate mass
cerebrum
   longitudinal fissure
   corpus callosum

lateral ventricle
third ventricle
corpus striatum (stri-AH-tum)
   caudate (KAW-dat) nuclei
   lentiform (LEN-tih-form) nucleus
      putamen (pyoo-TAH-men)
      globus pallidus (GLO-bus PAL-ih-dus)
cerebral cortex
   longitudinal fissure
   left/right hemisphere
   insula (IN-su-lah)
cerebral medulla

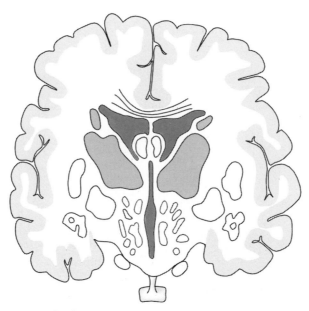

FIG. 15.5 Brain, frontal section through the hypophysis.

A FUNCTIONAL LOOK AT THE CEREBRUM

Cerebrum, Right Lateral View

Locate and *label* the cerebral regions in Figure 15.6. *Provide* a short description of the function of each area. Also *note* in which cerebral lobe the area is found.

primary visual area _____

_____

_____

general sensory area _____

_____

_____

primary motor area _____

_____

_____

premotor area _____

_____

_____

frontal eye field area _____

_____

_____

primary auditory (AW-dih-tory) area _____

_____

_____

primary gustatory (GUS-tah-to-re) area _____

_____

_____

motor speech (Broca's [BRO-kaz]) area _____

_____

_____

visual association area _____

_____

_____

somesthetic (so-mes-THET-ik) association area

_____

_____

_____

auditory association area _____

_____

_____

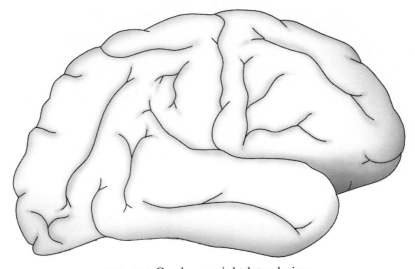

FIG. 15.6 Cerebrum, right lateral view.

CRANIAL NERVES

For each nerve below, *list* the following:

1. Number of the nerve.
2. Area and/or structure that is innervated.
3. Whether the nerve is sensory or mixed.
4. Foramen through which the nerve emerges from the cranium.

Locate and *label* the following in Figure 15.7.

olfactory (ol-FAK-to-re) nerve _____

_____

_____

optic (OP-tik) nerve _____

_____

_____

oculomotor (ok-u-lo-MO-ter) nerve _____

_____

_____

trochlear (TROK-le-ar) nerve _____

_____

_____

trigeminal (tri-JEM-ih-nal) nerve _____

_____

_____

abducens (ab-DOO-senz) nerve _____

_____

_____

facial (FA-shal) nerve _____

_____

_____

vestibulocochlear (ves-tib-u-lo-KOK-le-ar) nerve

_____

_____

_____

glossopharyngeal (glos-o-fah-RIN-je-al) nerve

_____

_____

_____

vagus (VA-gus) nerve _____

_____

_____

accessory (ak-SES-o-re) nerve _____

_____

_____

hypoglossal (hi-po-GLOS-al) nerve _____

_____

_____

Divisions of the Trigeminal Nerve

List the areas innervated by the following divisions of the trigeminal nerve. (*Obtain* this information from your textbook.)

ophthalmic (of-THAL-mik) branch _____

_____

_____

maxillary (MAX-sih-ler-e) branch _____

_____

superior alveolar (al-VE-o-lar) nerve _____

_____

mandibular (man-DIB-u-lar) branch _____

_____

_____

inferior alveolar nerve _____

_____

_____

CRITICAL THINKING

1. Which of the branches of the trigeminal would need to be anesthetized in dental work involving the teeth in the upper jaw? In the mandible?

   _____

   _____

   _____

   _____

2. Why does the lip become numb during dental work on the mandibular dentition?

   _____

   _____

   _____

   _____

   _____

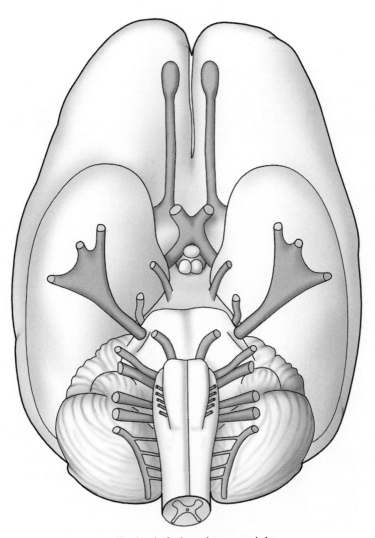

FIG. 15.7 Brain, inferior view, cranial nerves.

MAMMALIAN BRAIN DISSECTION

Rinse the sheep brain. When human materials are provided for demonstration, be sure to *correlate* these with your dissection.

Sheep Brain, Dorsal View

The cerebellum is deflected ventrad.

Locate the following on the sheep brain shown in Figure 15.8. Refer to CDA-14.

longitudinal fissure

left/right cerebral hemispheres

    frontal lobe

    parietal lobe

    occipital lobe

cerebellum

brain stem

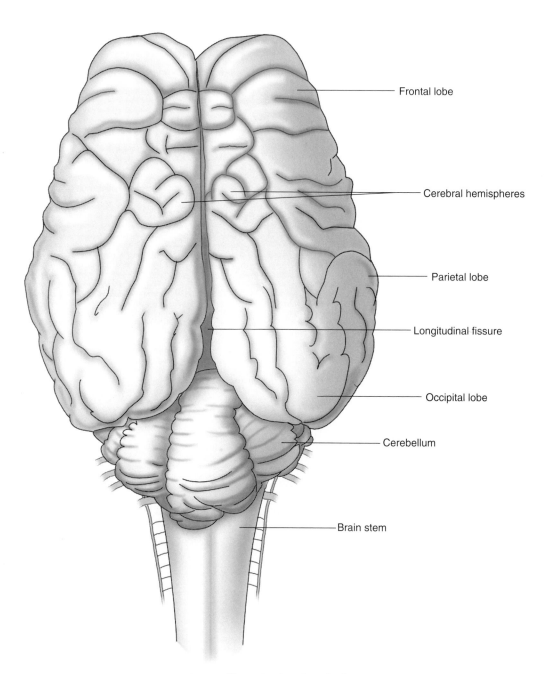

FIG. 15.8 Sheep brain, dorsal view.

Sheep Brain, Ventral View

Locate the following on the sheep brain shown in Figure 15.9. Refer to CDA-15.

cerebrum

    lateral sulcus

    temporal lobe of the cerebrum

diencephalon

    infundibulum

brain stem

    medulla oblongata

    pons

spinal cord

cranial nerves

    I. olfactory

    II. optic

    III. oculomotor

    IV. trochlear

    V. trigeminal

    VI. abducens

    VII. facial

    VIII. vestibulocochlear

    IX. glossopharyngeal

    X. vagus

    XI. accessory

    XII. hypoglossal

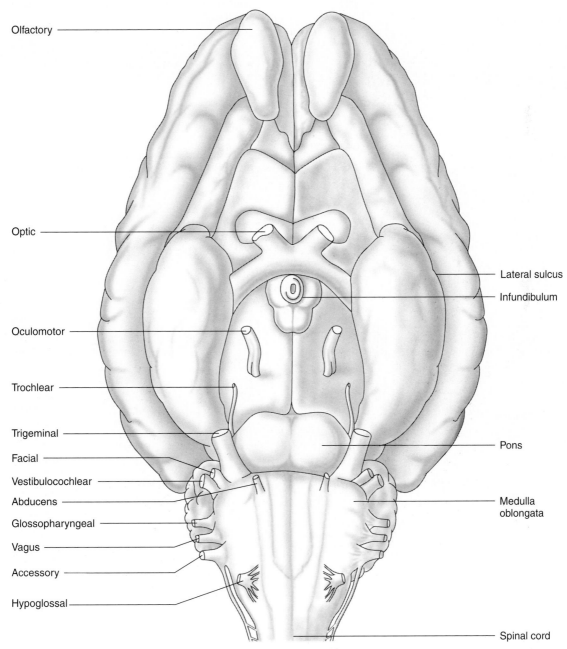

FIG. 15.9 Sheep brain, ventral view.

Sheep Brain, Midsagittal View, Right Hemisphere

Locate the following on the sheep brain shown in Figure 15.10. Refer to CDA-16.

lateral ventricle

third ventricle

    cerebral aqueduct

fourth ventricle

optic chiasma

corpus callosum

spinal cord

    central canal

brain stem

    medulla oblongata

    pons

    corpora quadrigemina

cerebellum

    arbor vitae

diencephalon

    pineal body

    thalamus

        intermediate mass

    hypothalamus

        infundibulum

    hypophysis (may be absent from your specimen)

Ask your instructor how to properly *store* or *dispose* of the specimen.

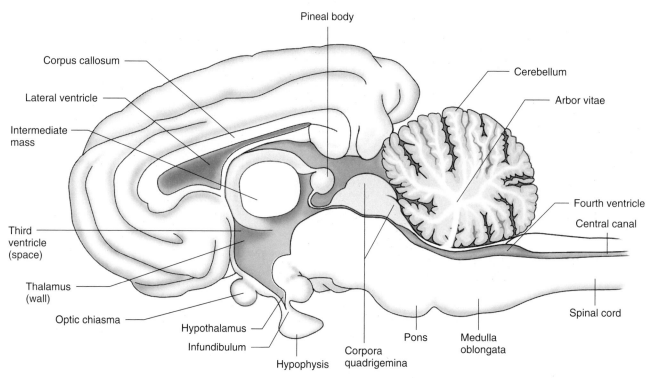

FIG. 15.10 Sheep brain, midsagittal view, right hemisphere.

THE SPINAL CORD AND SPINAL NERVES

The spinal cord is part of the central nervous system (CNS). In contrast to the brain, gray matter is deep and surrounds a central canal, and the white matter, myelinated axons of neurons, is located in the cortex of the spinal cord. The spinal cord carries motor and sensory neurons between the brain and the periphery. Another spinal cord function is the integration of peripheral reflexes. An example is the simple reflex of the knee jerk in response to a tap on the patellar tendon.

Spinal Cord, Cross Section

Locate and *label* the following in Figure 15.11.

gray matter

    anterior, lateral, and posterior horn

    gray commissure (KOM-ih-sh<u>u</u>r)

    central canal

white matter

    anterior, lateral, and posterior column

    anterior median fissure

    posterior median sulcus

spinal nerve (PNS)

    dorsal (posterior) root

       dorsal root ganglion (GANG-gl<u>e</u>-on)

    ventral (anterior) root

CRITICAL THINKING

1. *Define* the term tract.

    _____

    _____

    _____

    _____

2. What can be determined from the term lateral corticospinal tract?

    _____

    _____

    _____

    _____

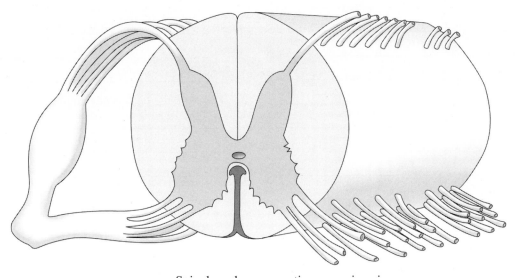

FIG. 15.11 Spinal cord, cross section, superior view.

3. *Define* the term ganglion.

_____

_____

_____

_____

4. What are the cerebral nuclei? *List* some functions of these.

_____

_____

_____

_____

5. *List* the cranial nerves by name, number, type (sensory, motor, visceral, somatic), innervation, and foramen by which each leaves the skull.

_____

_____

_____

_____

_____

_____

_____

_____

_____

_____

_____

_____

_____

_____

_____

_____

_____

_____

_____

_____

_____

_____

_____

_____

Spinal Cord and Spinal Nerves, Posterior View

Locate and *label* the following in Figure 15.12.

spinal nerve

    cervical

    thoracic

    lumbar

    sacral

    coccygeal

brachial plexus (PLEX-sus)

lumbar plexus

sacral plexus

conus medularis (KO-nus med-u-LAR-is)

cauda equina (KAW-daw ek-KWI-nah)

CRITICAL THINKING

1. Which spinal nerves form the following?

    a. brachial plexus

_____

_____

_____

_____

    b. lumbar plexus

_____

_____

_____

_____

    c. sacral plexus

_____

_____

_____

_____

2. *Translate* the term cauda equina.

_____

_____

_____

_____

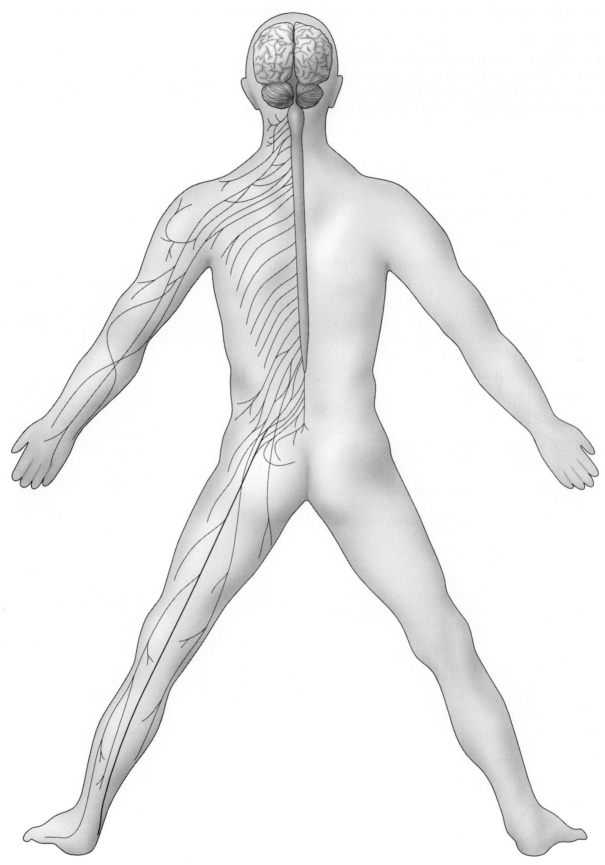

FIG. 15.12 Spinal cord and nerves, posterior view.

THE SPINAL CORD AND SPINAL NERVES

Distribution of Spinal Nerve Function

The following information is generally not part of a course in introductory anatomy. It is provided here as a reference. Figure 15.13 shows the dermatomes (DER-mah-t<u>o</u>mz)—regions of the skin innervated by specific spinal nerves. Note the segmental arrangement of the dermatomes. By knowing which spinal nerve innervates a particular dermatome, you may determine what segment of the spinal cord or spinal nerve is malfunctioning. That is, if a dermatome is stimulated (scratched or poked) and the patient does not perceive the sensation, the nerve supplying that dermatome is implicated.

FUNCTIONS OF THE SPINAL NERVES

Spinal Nerve Group	Name	Function
C_1–C_4 ventral rami	**cervical plexus**	To skin and muscles of the head and neck, superior shoulders and diaphragm by the **phrenic** nerves.
dorsal rami		To dermatomes (see Figure 15.13).
C_5–C_8 and T_1 ventral rami	**brachial plexus**	To superior appendages and shoulders. The deltoid and teres minor muscle are supplied by the **axillary nerve**, the arm and forearm flexors by the **musculocutaneous nerve**, the arm and forearm extensors by the **radial nerve**, the muscles of the anterior arm and palm by the **median nerve**, and the medial muscles of the forearm and palm by the **ulnar nerve**.
dorsal rami		To dermatomes (see Figure 15.13).

FUNCTIONS OF THE SPINAL NERVES (continued)

Spinal Nerve Group	Name	Function
T_2–T_{11} ventral rami	**intercostal nerves**	No plexus is formed by these nerves. The ventral rami supply the intercostal muscles and skin of the chest wall as well as abdominal musculature and skin. The dorsal rami supply the deep back musculature and skin of the back.
dorsal rami		Dermatomes of dorsal thorax and deep back musculature (see Figure 15.13).
T_{12} ventral ramus	**subcostal nerve**	Inferior rectus abdominis muscle and overlying skin.
L_1–L_4 ventral rami	**lumbar plexus** (L_4 also contributes to the sacral plexus)	To anteriolateral abdominal wall, external genitalia, inferior appendage. Like the brachial plexus, this plexus forms several nerves, the largest being the **femoral nerve** supplying the flexors of the thigh, extensors of the leg and the skin of the medial and anterior thigh and medial leg and foot.
dorsal rami		Dermatomes (see Figure 15.13).
L_4–L_5 and S_1–S_4 ventral rami	**sacral plexus**	Supplies the buttocks, perineum and inferior appendages. The most prominent nerve is the **sciatic nerve** that serves all the muscles of the leg and foot.
dorsal rami		The dermatomes of the perianal region and inferior appendages (see Figure 15.13).

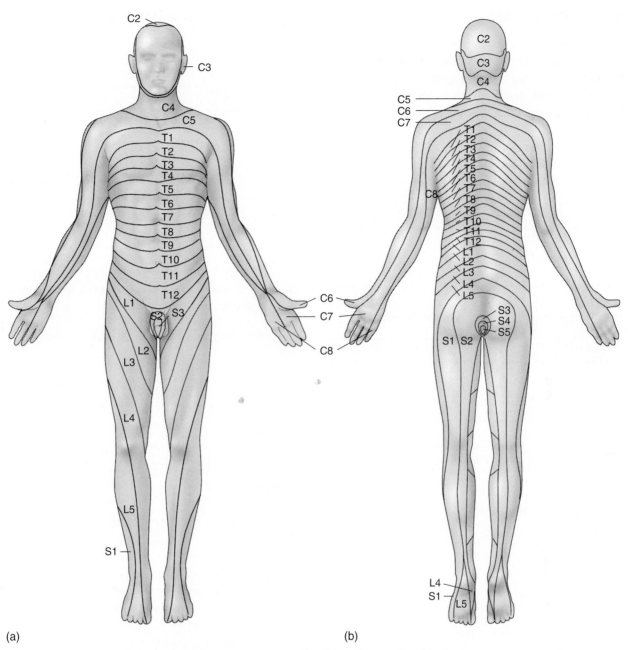

FIG. 15.13 Dermatomes, anterior (a) and posterior (b) views.

Spinal Reflex Arc

A reflex arc consists of at least two neurons that conduct information from a peripheral receptor to the central nervous system. In the central nervous system (in this example the spinal cord) a motor neuron is stimulated and the impulse is conducted to an effector.

Define the following components of the reflex arc and *locate* each on Figure 15.14.

receptor _____

_____

sensory neuron _____

_____

center (may contain an association or internuncial neuron) _____

_____

motor neuron _____

_____

effector _____

_____

CRITICAL THINKING

1. Name a somatic reflex.

    _____

    _____

    _____

    _____

2. Name a visceral reflex.

    _____

    _____

    _____

    _____

3. What is a dermatome?

    _____

    _____

    _____

    _____

4. What symptoms would injury to the femoral nerve produce?

    _____

    _____

    _____

    _____

5. Why does the use of a crutch often produce "crutch palsy"?

    _____

    _____

    _____

    _____

6. Where would the spinal cord have to be damaged to produce quadriplegia?

    _____

    _____

    _____

    _____

AUTONOMIC NERVOUS SYSTEM

The **autonomic nervous system (ANS)** is a visceral motor system. It consists of visceral efferent neurons that regulate the smooth muscles (of viscera, blood vessels, intrinsic muscles of the eye, etc.), heart, and activities of the glands, all of which are generally "involuntary"—out of conscious control. These regulatory neurons originate in the central nervous system and pass (via two serial neurons) to a visceral effector. The first neuron (the **preganglionic** (pre-gang-gle-ON-ik) **neuron**) in each autonomic path extends from the central nervous system to a peripheral ganglion where it synapses with a second neuron (the **postganglionic neuron**) extending to the effector. The postganglionic neuron secretes either **acetylcholine** (as-eh-til-KO-len) or **norepinephrine** (nor-ep-ih-NEF-rin) as neurotransmitters. If the neurotransmitter is norepinephrine, then the pathway is termed **sympathetic** (sim-pa-THET-ik). If the neurotransmitter is acetylcholine, the pathway is called **parasympathetic** (par-a-sim-pa-THET-ik). The two divisions of the autonomic nervous system often have an antagonistic influence on the affected organ.

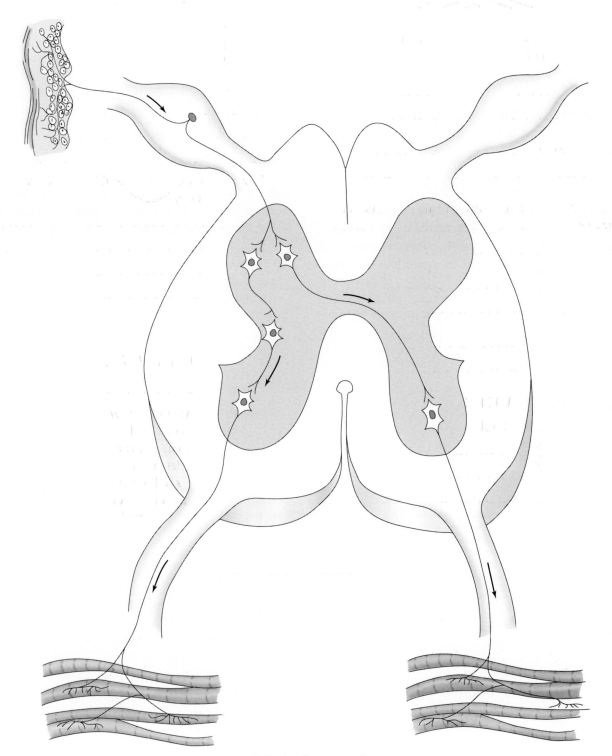

FIG. 15.14 Spinal reflex arc, schematic.

Most visceral organs have this dual innervation so that one division may be excitatory while the other acts as an inhibitor.

Note how the description of the visceral motor neurons above differs from that of somatic motor neurons. Somatic motor neurons have one neuron in the pathway from the central nervous system to the peripheral effector (skeletal muscle). In somatic motor pathways there is only one type of neuron and it secretes acetylcholine as the neurotransmitter. This type of neuron is always excitatory.

Anatomy of the Autonomic Nervous System

Locate the components on Figure 15.15 as they are discussed.

The preganglionic neurons of the sympathetic pathways originate and have their cell bodies in the lateral horns of the thoracic and the two superior lumbar segments of the spinal cord. This division is thus also called the **thoracolumbar** (tho-ra-ko-LUM-bar) **division**. The origins of the preganglionic neurons of the parasympathetic pathways are quite different. These begin and have their cell bodies in the oculomotor, facial, glossopharyngeal, and vagal nuclei of the brain stem. A few parasympathetics originate in the lateral horns of sacral segments 2, 3, and 4. The parasympathetic division is also called the craniosacral (kra-ne-o-SA-kral) division.

The synapses between the pre- and postganglionic neurons occur in the **autonomic ganglia**. There are three types of ganglia:

1. **paravertebral** (par-ah-VER-teh-bral) or **lateral ganglia**

2. **prevertebral** (pre-VER-teh-bral) or **collateral ganglia**

3. **terminal** or **intramural ganglia**

Paravertebral ganglia, as the name implies, lie in a column that is parallel to the vertebrae. Part of the sympathetic division has synapses in these ganglia. Prevertebral ganglia are anterior to the vertebrae in the region of the abdominal aorta. The rest of the sympathetic division has synapses in these ganglia. Terminal ganglia are located at the visceral effector and are the site of the synapses in the parasympathetic division. Finally, the postganglionic neurons begin at the autonomic ganglia, in all three types, and pass to the effector.

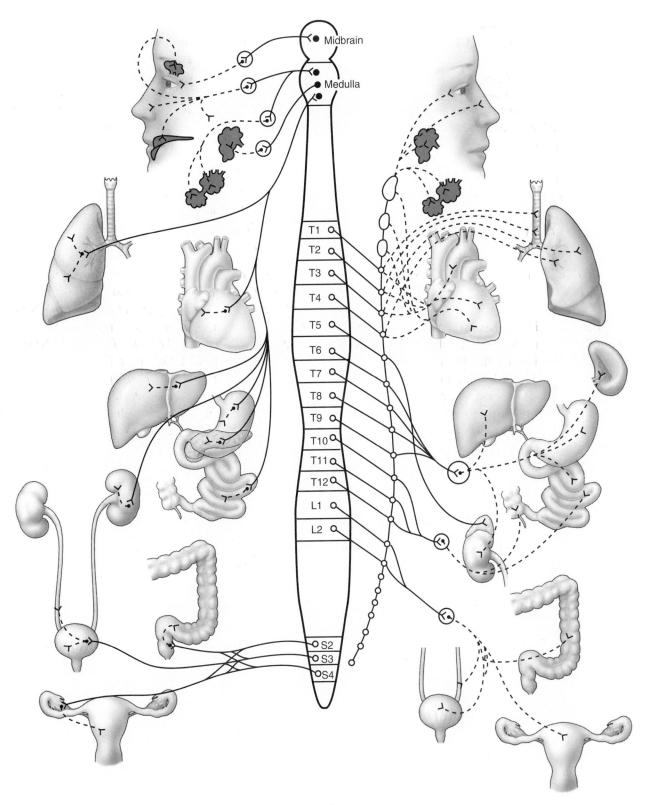

FIG. 15.15 Autonomic nervous system.

SHORT ANSWER

1. *Trace* a neuronal pathway from a pain receptor in the foot to the primary sensory cortex. In the order of occurrence, *list* the sites of the synapses, the rami and roots of spinal nerves, the column of the spinal cord, the brain divisions involved, and the location of the translation site in the brain.

2. *Trace* a neuronal pathway from the primary motor cortex to the intercostal muscles. In the order of occurrence, *list* the sites of the synapses, the rami and roots of spinal nerves, the column of the spinal cord, the brain divisions involved, and the location of the control center in the brain.

3. *Trace* a parasympathetic pathway from the medulla oblongata to the smooth muscle of the stomach. In the order of occurrence, *list* the sites of the synapses, the pathway followed from the brain to the organ, and the brain divisions involved.

4. *Trace* a parasympathetic pathway from the medulla oblongata to an organ innervated by a sacral spinal nerve. In the order of occurrence, *list* the sites of the synapses, the pathway followed from the brain to the organ, and the brain divisions involved.

5. *Trace* a sympathetic pathway from the medulla oblongata to an organ innervated by a thoracic spinal nerve. In the order of occurrence, *list* the sites of the synapses, the pathway followed from the brain to the organ, and the brain divisions involved.

MATCHING

Match the cerebral regions in column A with their functions or descriptions in column B.

A	B
_____ 1. precentral gyrus	a. learned motor area
_____ 2. medial temporal lobe	b. motor speech area
_____ 3. postcentral gyrus	c. primary olfactory area
_____ 4. lateral temporal lobe	d. primary visual area
_____ 5. Broca's area	e. primary auditory area
_____ 6. occipital lobe	f. primary motor area
_____ 7. premotor area	g. primary sensory area

Match the descriptions in column A with the terms relating to the neuron in column B.

A	B
_____ 1. norepinephrine	a. axon
_____ 2. produces myelin	b. dendrite
_____ 3. short, numerous processes	c. cell body
_____ 4. long process with synaptic bulbs	d. myelin
_____ 5. contains the nucleus	e. neurolemmocytes
_____ 6. specific site of synaptic vesicle	f. neurotransmitter
_____ 7. increases speed of conduction	g. synapse
_____ 8. space separating two neurons	h. synaptic bulb
_____ 9. bulbs, cleft, vesicles, transmitter substances	i. synaptic cleft

Match the descriptions in column A with the brain regions in column B.

A	B
_____ 1. visual and auditory reflexes	a. cerebellum
_____ 2. secondary respiratory center	b. cerebral cortex
_____ 3. connects cerebral hemispheres	c. corpora quadrigemina
_____ 4. visceral control (breathing, heart, coughing)	d. corpus callosum
_____ 5. primitive drives, emotions, releasing factors	e. hypothalamus
_____ 6. conscious centers	f. medulla oblongata
_____ 7. integrates somatic motor impulses and proprioceptors	g. pons

MULTIPLE CHOICE

_____ 1. Which of the following operates the triceps muscle?

    a. autonomic neuron
    b. visceral motor neuron
    c. somatic motor neuron
    d. somatic sensory neuron

_____ 2. An axon differs from a dendrite in that

    a. it has synaptic vesicles.
    b. it is always longer.
    c. it alone can depolarize.
    d. the dendrite contains the nucleus.

_____ 3. Which is false regarding the spinal cord?

    a. Nuclei, groups of neuron cell bodies, and dendrites are found in the gray matter.
    b. The ventral horns of gray matter give rise to the motor roots.
    c. The central canal contains cerebrospinal fluid.
    d. White matter consists of myelinated dendrites of sensory neurons.

_____ 4. Which is NOT an example of a visceral effector in the ANS?

    a. skeletal muscle   c. gland
    b. visceral muscle   d. cardiac muscle

_____ 5. Assuming a drop of cerebrospinal fluid is formed in a lateral ventricle, which space is encountered next?

    a. fourth ventricle
    b. third ventricle
    c. subarachnoid space
    d. cerebral aqueduct

_____ 6. A neuron that has a single axon and several dendrites is called

    a. afferent.   d. multipolar.
    b. bipolar.   e. axoaxonic.
    c. unipolar.

_____ 7. Which of the following innervates the muscles involved in walking?

    a. visceral motor neuron
    b. somatic motor neuron
    c. visceral sensory neuron
    d. somatic sensory neuron
    e. interneuron

_____ 8. An effector is

    a. the region in the brain in which a nervous impulse originates.
    b. the organ on which a motor neuron ends.
    c. a particular type of motor neuron.
    d. a particular type of sensory neuron.
    e. described by more than one of the above.

_____ 9. An interneuron is

    a. sensory.
    b. motor.
    c. found in the brain only.
    d. a neuron that integrates sensory and motor activities.
    e. found in the spinal cord only.

_____ 10. A neurotransmitter
   a. is released from the synaptic vesicle.
   b. bridges the synaptic cleft.
   c. produces depolarization in the postsynaptic neuron.
   d. is a hormone such as dopamine.
   e. is described by all of the above.

_____ 11. Place the following in the correct order, begin in the CNS:
   1. neuromuscular junction at heart muscle
   2. sympathetic column in spinal cord
   3. cardiac center in medulla oblongata
   4. ventral root of thoracic spinal nerve
   5. paravertebral ganglion of sympathetic trunk, synapse
   6. preganglionic neuron
   7. ganglionic neuron

   a. 1, 2, 3, 4, 5, 6, 7
   b. 3, 2, 6, 4, 5, 7, 1
   c. 3, 6, 2, 7, 5, 4, 1
   d. 1, 4, 2, 5, 6, 7, 3
   e. 2, 3, 7, 4, 5, 1, 6

_____ 12. From the following list of characteristics, select those that are correct pertaining to some or all spinal nerves.
   1. four pairs of coccygeal nerves
   2. eight pairs of cervical nerves
   3. may contain sensory pathways
   4. may contain somatic motor pathways
   5. may contain visceral motor pathways
   6. formed from two roots
   7. 10 pairs of lumbar nerves
   8. ventral root has ganglion
   9. may contain parasympathetic pathways
   10. leave CNS via transverse foramina

   a. 2, 3, 4, 5, 6, 9
   b. 2, 4, 5, 6, 7, 8
   c. 2, 3, 4, 7, 8, 9, 10
   d. 1, 3, 6, 7, 8, 10
   e. none of the above

_____ 13. Which of the following is false regarding the spinal cord?
   a. Nuclei, groups of neuron cell bodies and dendrites, are found in the gray matter.
   b. The anterior horns of gray matter give rise to the motor roots.
   c. The central canal contains CSF.
   d. The dorsal horns receive the sensory roots.
   e. White matter consists of myelinated dendrites and cell bodies of sensory neurons.

_____ 14. Assuming CSF originates only in the lateral ventricles, select the correct sequence of formation, circulation, and reabsorption of CSF from the list below.
   1. third ventricle
   2. lateral ventricle
   3. arachnoid villi
   4. choroid plexus
   5. superior sagittal sinus
   6. fourth ventricle
   7. cerebral aqueduct
   8. subarachnoid space

   a. 1, 2, 3, 4, 5, 6, 7, 8
   b. 4, 1, 2, 3, 8, 6, 7, 5
   c. 3, 2, 1, 5, 6, 7, 8, 4
   d. 4, 2, 1, 7, 6, 8, 3, 5
   e. none of the sequences

16 Sensation

Contents

Objectives

1. List the types of receptors by location.
2. List the types of receptors by type of stimulus detected.
3. Name and describe the function of the accessory structures of the eye.
4. Name and describe the function of the parts of the eye.
5. List and define six common eye problems.
6. Name and describe the function of the parts of the ear.

FUNCTIONS OF A RECEPTOR

Before the brain reports a sensation, a receptor or sense organ must first detect the stimulus and transduce it to a nerve impulse. A sensory pathway must then conduct the impulse to the central nervous system for translation. Depending on the type of pathway, the impulse may be processed and perceived at a conscious level (such as vision) or at a subconscious level (such as heart rate).

TYPES OF RECEPTORS

Receptors are organized by location and by the type of stimulus they detect.

There are three receptor types organized by location:

Exteroceptors (eks-ter-o-SEP-tors)

Visceroceptors (vis-er-o-SEP-tors)

Proprioceptors (pro-pre-o-SEP-tors)

The **exteroceptors** detect the stimuli for vision, olfaction, audition, gustation, touch, pressure, temperature, and pain. Exteroceptors provide information about the external environment.

The **visceroceptors** provide information about the internal "housekeeping" of the body. These receptors are located in the viscera, blood vessels, and hypothalamus.

The **proprioceptors** provide information about movement and position of the skeleton and skeletal muscles in relation to the external environment and in relation to other skeletal and muscle elements.

The receptor types organized by stimulus detected are as follows:

Mechanoreceptors—detect mechanical deformation such as pressure and stretching due to touch, vibration, proprioception, audition, equilibrium, and blood pressure.

Thermoreceptors—are sensitive to changes in temperature.

Photoreceptors—detect light.

Chemoreceptors—are sensitive to chemicals and are responsible for gustation, olfaction, and detection of chemicals in body fluids.

Nociceptors (no-se-SEP-tors)—detect pain due to tissue damage.

In the exercises that follow, we will concentrate on two of the exteroceptors as examples of specialized receptors. The sensations are vision and audition. The receptors are the eye and the ear. Your instructor may discuss other specialized receptors with you. *Consult* your textbook for more information on those.

THE EYE

Eye, Frontal and Midsagittal Views

Locate and *label* the following in Figure 16.1(a) and (b).

accessory structures

    eyebrow

    eyelid

       tarsal (TAR-sl) (Meibomian) gland

    conjunctiva (kon-JUNK-tih-vah)

    eyelash

    lacrimal (LAK-rih-mal) apparatus

       lacrimal gland

       lacrimal duct

       nasolacrimal duct

    extrinsic (eks-TRIN-sik) muscles

       superior oblique (o-BLEK)

       inferior oblique

       superior rectus (REK-tus)

       inferior rectus

       medial rectus

       lateral rectus

eyeball

    fibrous tunic (TOO-nik)

       sclera (SKLE-rah)

       cornea (KOR-ne-ah)

    vascular tunic

       choroid (KO-royd)

       ciliary (SIL-e-er-e) body

       iris (I-ris)

          pupil (PYOO-pil)

    nervous tunic or retina

       rods (not illustrated)

       cones (not illustrated)

       fovea (FO-ve-ah)

       optic disc

    lens

       suspensory ligament

    anterior cavity

       anterior chamber

       posterior chamber

       aqueous (AH-kwe-us) humor

    posterior cavity

       vitreous (VIT-re-us) humor

MAMMALIAN EYE DISSECTION

Rinse the eye. *Locate* the following external features on the intact eye:

oblique muscles

rectus muscles

optic nerve (white in color with the ophthalmic artery next to it)

cornea

sclera

iris

pupil

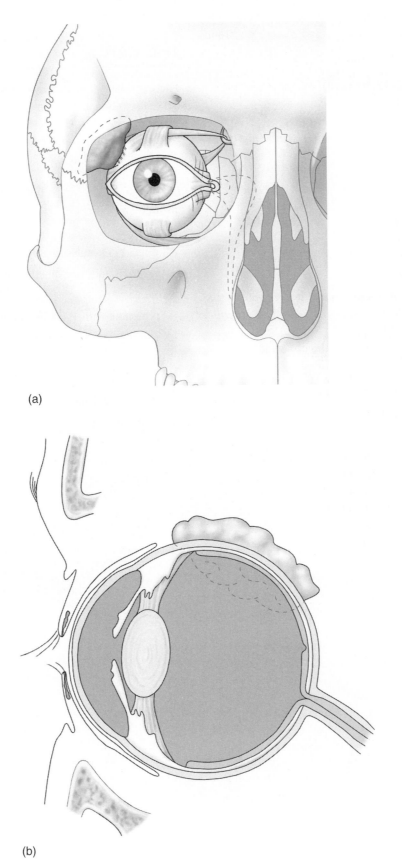

(a)

(b)

FIG. 16.1 Eye and accessory structures, anterior (a) and midsagittal (b) views.

MAMMALIAN EYE DISSECTION

With a razor blade *cut* a small opening through the wall of the eyeball posterior to the cornea (Figure 16.2). *Continue* this cut with your scissors parallel to and one-half of a centimeter behind the cornea and *remove* the front of the eyeball.

⚠ DO THIS CAREFULLY SO AS NOT TO DISTURB THE INTERNAL STRUCTURES.

Locate the following internal features on the dissected eye:

retina

optic disk

choroid

suspensory ligaments

lens

Remove the lens (if it is clear, *place* it over some fine print and *note* the effect). Also, *cut* into the lens and *observe* the layered detail.

Locate the following features on the dissected eye.

posterior cavity

anterior cavity

    posterior chamber

    anterior chamber

iris

Ask your instructor how to properly store or dispose of the specimen.

THE EAR

The ear consists of the **outer ear**, **middle ear**, and **inner ear**. While the outer ear and middle ear serve to support the aural function, the inner ear contains the receptors for both sound (the cochlea) and equilibrium (the vestibule and semicircular canals).

Locate and *label* the following in Figure 16.3(a) through (c).

external ear

    pinna (PIN-nah) or auricle (AW-re̱-kl)

    external auditory meatus (me̱-A̱-tus)

middle ear

    tympanic (tim-PAN-ik) membrane

    auditory ossicles (OS-sih-kls) [16-3(b)]

    malleus (MAL-e̱-us)

    incus (ING-kus)

    stapes (STA̱-pe̱z)

    oval window

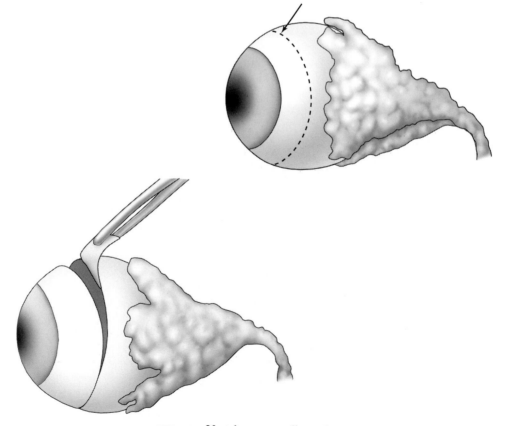

FIG. 16.2 Vertebrate eye dissection.

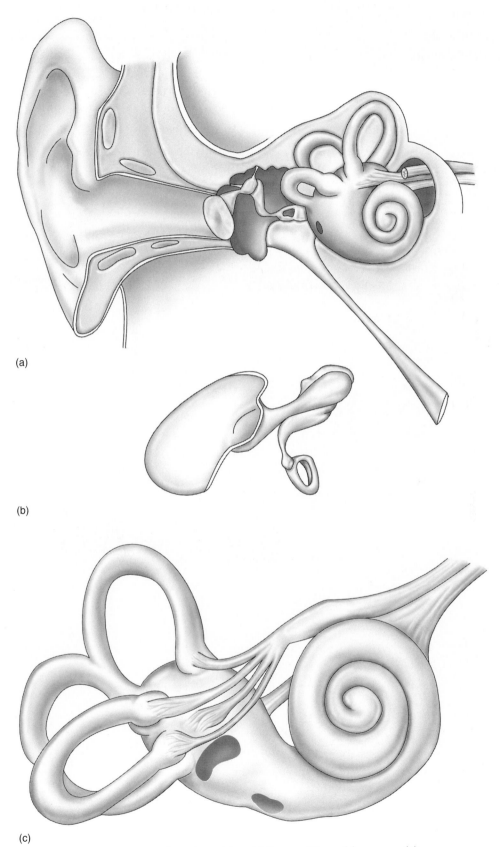

(a)

(b)

(c)

FIG. 16.3 Ear, all regions (a), middle ear (b), and inner ear (c).

internal ear [16-3(c)]

  vestibulocochlear (ves-tib-u-lo-KOK-le-ar) nerve

  vestibular branch

  cochlear branch

  cochlea (KOK-le-ah)

  vestibule (VES-tih-bul)

  round window

  semicircular canals

On Figure 16.3(c), *draw* the outline of the membranous labyrinth using dotted lines. Once finished, *label* the following:

utricle (U-tre-kl)

saccule (SAK-ul)

ampullae (am-PUL-le)

Ear, Cochlear Cross Section

Locate and *label* the following in Figure 16.4.

scala vestibuli (SKA-lah VES-tih-bu-li)

scala tympani (tim-PA-ni)

vestibular membrane

basilar (BAS-ih-lar) membrane

cochlear duct

  tectorial (tek-TO-re-al) membrane

  hair cells

cochlear branch of the vestibulocochlear nerve (VIII)

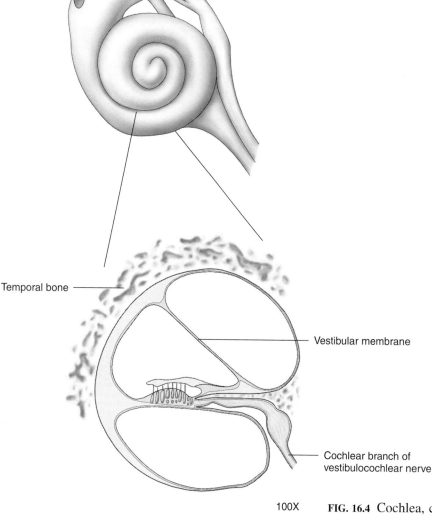

Temporal bone

Vestibular membrane

Cochlear branch of vestibulocochlear nerve

100X

FIG. 16.4 Cochlea, cross section.

SHORT ANSWER

1. What is a cataract?

2. What is glaucoma?

3. What is conjunctivitis?

4. *Define* lacrimate.

5. *Define* myopia and *describe* the cause.

6. *Define* hyperopia and *describe* the cause.

7. *Define* presbyopia and *describe* the cause.

MATCHING

Match the descriptions in column A with the terms in column B.

A

_____ 1. lens portion of the fibrous tunic

_____ 2. contains the photoreceptors

_____ 3. cornea is too flat

_____ 4. pigmented region

_____ 5. highest visual acuity in daylight

_____ 6. loss of elasticity of the accommodating lens

_____ 7. blind spot

_____ 8. accommodation

_____ 9. supports the accommodating lens

_____ 10. fills the posterior cavity

B

a. choroid

b. suspensory ligaments

c. cornea

d. fovea centralis

e. hyperopia

f. accommodating lens

g. optic disk

h. presbyopia

i. retina

j. vitreous humor

MULTIPLE CHOICE

_____ 1. Which of the following describes the cochlea?

    a. mechanoreceptor   d. for hearing
    b. exteroceptor   e. all of the above
    c. special sense

_____ 2. Which of the following is a special sense?

    a. touch   d. pressure
    b. proprioception   e. gustation
    c. pain

_____ 3. Which of the following includes pain receptors?

    a. mechanoreceptors   d. chemoreceptor
    b. nociceptors   e. none of the above
    c. photoreceptor

_____ 4. The process of adjusting for objects near the eye is

    a. presbyopia.   d. accommodation.
    b. hyperopia.   e. glaucoma.
    c. myopia.

_____ 5. Which of the following is NOT associated with the aqueous humor?

    a. anterior cavity   d. glaucoma
    b. choroid plexus   e. optic nerve
    c. ciliary body

_____ 6. Which of the following is NOT associated with the rods?

    a. black and white vision
    b. cornea
    c. optic nerve
    d. bipolar neuron
    e. nervous tunic

_____ 7. Which of the following is a receptor for static equilibrium?

    a. cochlea   d. macula
    b. retina   e. ampulla
    c. cupula

17 Endocrine System

Contents

17

Objectives

1. Define an endocrine gland.
2. List the major endocrine glands.
3. Describe the locations of the endocrine glands.
4. List the hormones of each endocrine gland.
5. List the target organ of each of the hormones.
6. Describe the effects of each endocrine hormone.

FUNCTIONS OF THE ENDOCRINE SYSTEM

The **endocrine** (EN-do-krin) **system** consists of numerous glands found throughout the body. These glands secrete a large variety of **hormones** (HOR-monz) that act as chemical messengers controlling many bodily functions. Some of these functions are the regulation of reproductive cycles, body temperature, concentration of ions in the blood, blood sugar, blood volume, growth, development, birth and lactation, and stress management. Together with the nervous system, the endocrine system functions to maintain homeostasis. The study of the endocrine system is **endocrinology** (en-do-kri-NOL-o-je).

LOCATION OF THE ENDOCRINE GLANDS

Locate and *label* the following in Figure 17.1.

adrenal (ah-DRE-nal)

hypophysis (hi-POF-ih-sis)

hypothalamus (hi-po-THAL-ah-mus)

ovary (O-vah-re)

pancreas (PAN-kre-as)

parathyroid (par-ah-THI-royd)

pineal (PIN-e-al)

testis (TES-tis)

thymus (THI-mus)

thyroid (THI-royd)

All of the endocrine glands secrete hormones directly into the bloodstream. From there the hormones travel by the general circulation to the target organ where they exert an effect by interaction with the plasma membrane or with the genome of the cell.

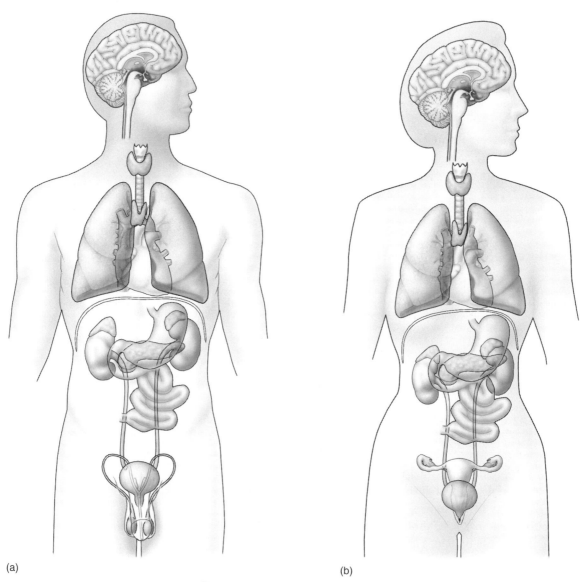

(a)

(b)

FIG. 17.1 Endocrine glands, male (a) and female (b).

HORMONES, TARGET ORGANS, AND EFFECTS OF THE ENDOCRINE GLANDS

After each of the glands listed below, *name* the hormones produced, an abbreviation (if appropriate) for the hormone, the target organ, and the effect on the target organ. An example (thyrotropin) is given.

Gland	Hormone(s)	Target Organ	Effect
adrenal medulla			
adrenal cortex			
hypophysis	**thyrotropin (TSH)**	**thyroid gland**	**Controls the secretion and synthesis of thyroid hormones by the thyroid gland.**
hypothalamus			
ovary			
pancreas			

Gland	Hormone(s)	Target Organ	Effect
parathyroid			
pineal			
testis			
thymus			
thyroid			

SHORT ANSWER

1. *Describe* the anatomical symptoms of acromegaly.

2. *Explain* the difference between an exocrine and endocrine gland.

3. Why would a complete adrenalectomy require two incisions?

MATCHING

Match the glands in column A with the hormones in column B.

A	B
_____ 1. anterior pituitary	a. adrenocorticotropic hormone
_____ 2. posterior pituitary	b. calcitonin
_____ 3. hypothalamus	c. estrogen
_____ 4. thyroid	d. insulin
_____ 5. pancreas	e. oxytocin
_____ 6. parathyroid	f. parahormone
_____ 7. ovary	g. releasing hormone

18 Respiratory System

Contents

Objectives

1. Describe the anatomy and functions of the nasal cavity.

2. Describe and name the components of the pharynx.

3. Describe and name the cartilages of the larynx.

4. Describe the anatomy of the trachea and its divisions.

5. Describe the gas exchange surfaces.

6. Name the skeletal muscles involved in breathing.

7. Name and describe the location of the membranes of the thoracic cavity.

8. Trace the pathway of inhaled air from the external nares to the alveoli.

FUNCTIONS OF THE RESPIRATORY SYSTEM

The respiratory system (breathing system) functions to exchange gases between the air and blood. Oxygen diffuses from the lung to the capillary lumen where it is carried to the tissues by the red blood cells. Carbon dioxide leaves the blood and is exhaled. Oxygen is used by the mitochondria of the cells during cellular respiration. Carbon dioxide is produced as a waste product of cellular metabolism.

RESPIRATORY TRACTS DEFINED

The **upper respiratory tract** consists of the nose, nasal cavities, nasopharynx, oropharynx, and laryngopharynx.

The **lower respiratory tract** consists of the larynx, trachea, and bronchi of the lung proper.

UPPER RESPIRATORY TRACT

Head and Neck: Sagittal Section

Locate and *label* the following in Figure 18.1.

*external nares (N<u>A</u>-r<u>e</u>z)

  nasal cavity

    conchae (KONG-k<u>e</u>)

      superior

      middle

      inferior

    meatus (m<u>e</u>-<u>A</u>-tus)

      superior

      middle

      inferior

    septum (SEP-tum)

  paranasal sinus (S<u>I</u>-nus)

    frontal

    ethmoid

    sphenoid

    maxillary

  nasopharynx (n<u>a</u>-s<u>o</u>-FAR-inks)

    auditory tube

      auditory tube orifice (<u>O</u>R-ih-fis)

    pharyngeal tonsil (TON-sil) or adenoid (AD-eh-n<u>oy</u>d)

*oropharynx (<u>o</u>-r<u>o</u>-FAR-inks)

    palatine (PAL-ah-t<u>in</u>) tonsil

  laryngopharynx (lah-ring-g<u>o</u>-FAR-inks)

  trachea (TR<u>A</u>-k<u>e</u>-ah)

LARYNX

The larynx (L<u>A</u>R-inks) is the "voice box" or "Adam's apple." It functions first as a valve to prevent solids and liquids from entering the lower respiratory tract and second as a sound-producing organ. The structure consists of several unpaired and several paired cartilages and the musculature necessary to operate the larynx.

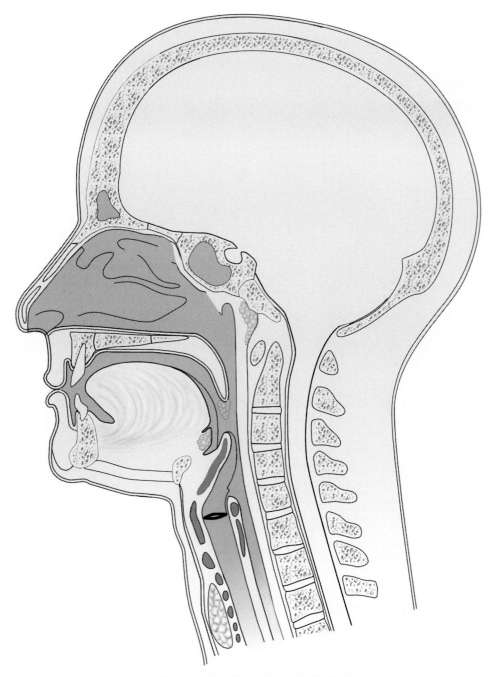

FIG. 18.1 Head and neck, right sagittal view.

Cartilages of the Larynx

Locate and *label* the following in Figure 18.2(a) through (d).

*thyroid cartilage

*cricoid (KR<u>I</u>-k<u>o</u>yd) cartilage

epiglottis (ep-ih-GLOT-is)

arytenoid (ar-eh-TEH-n<u>o</u>yd) cartilage

vocal folds (see again Figure 18.1)

    ventricular fold

    true vocal fold

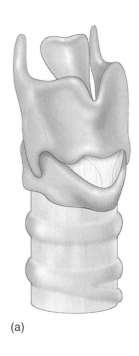

(a)

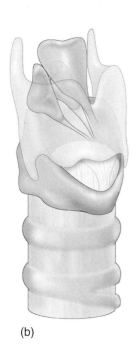

(b)

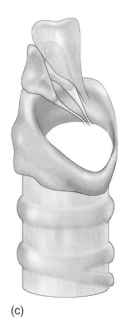

(c)

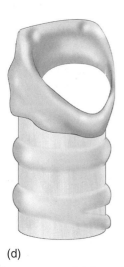

(d)

FIG. 18.2 Larynx, anterior view (a), transparent perspective (b), thyroid cartilage removed (c), and thyroid, epiglottal, arytenoid cartilages removed (d).

Lower Respiratory Tract: Anterior View with Ribs And Sternum

Locate and *label* the following in Figure 18.3(a).

larynx

trachea

    tracheal cartilage

primary bronchi (BRONG-kī)

left lung

right lung

diaphragm (DĪ-ah-fram) (dashed line)

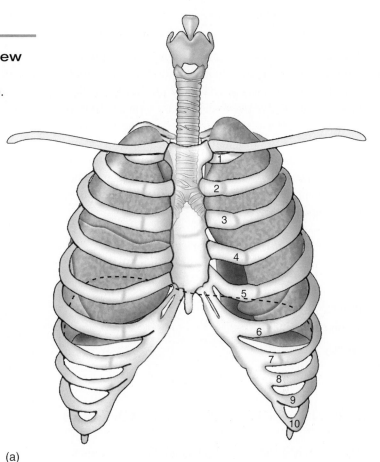

(a)

Lower Respiratory Tract: Schematic, Anterior View

Locate and *label* the following in Figure 18.3(b).

thoracic cavity

pleural (PLOO-ral) membranes

    parietal pleura (pah-RĪ-eh-tal PLOO-rah)

    visceral pleura

pleural cavity

mediastinum (mē-dē-as-TĪ-num)

diaphragm

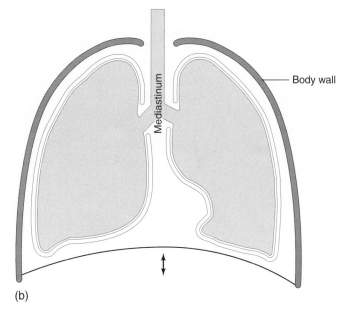

(b)

FIG. 18.3 Lower respiratory tract, anterior view (a) and schematic, anterior view (b).

RESPIRATORY TREE

The trachea divides to form the right and left primary bronchi. These in turn divide to form the secondary bronchi (two on the left and three on the right). Secondary bronchi give rise to the tertiary bronchi (10 right and 10 left). These supply the lobes of the lung (two left, three right) and the tertiary bronchi supply regions of lung tissue called bronchopulmonary (brong-ko-PUL-mo-ner-e) segments. The appearance of the trachea and bronchi, when viewed inverted, is the origin of the term **respiratory tree.**

Respiratory Tree, Anterior View

Locate and *label* the following in Figure 18.4(a).

trachea

primary bronchi

secondary bronchi

tertiary bronchi

BRONCHOPULMONARY SEGMENTS

Figure 18.4(b) shows the bronchopulmonary segments supplied by the tertiary bronchi.

CAT DISSECTION

If you have not yet opened the thoracic and abdominal cavities of the cat, do so now. *Turn* to the section on opening the body cavity in Chapter 13. *Refer* to CDA-17.

At the bottom of the thoracic cavity, *locate* the **diaphragm**, a broad muscular membrane. Cranial to this find the **left** and **right lungs** and note the **mediastinum**, the space between them that includes the heart and other organs. By pulling the lungs away from the heart you will *notice* the **primary bronchi** branching from the **trachea**. At the cranial end of the trachea *find* the **larynx** and note that the anterior surface of this is primarily composed of the **thyroid cartilage**. Now *make* a ventral midsagittal cut along the trachea and through the larynx. *Open* the incision and *find* the cut edges of the thyroid cartilage. Just caudal to this is a smaller cartilage, the **cricoid cartilage**. Inside this, the **mucous membrane** is thrown into a lateral pair of folds called the **vocal folds**. *Locate* the **epiglottis**, which is a soft cartilage at the cranial end of the larynx. Finally *find* the openings to the **esophagus** and **oropharynx**.

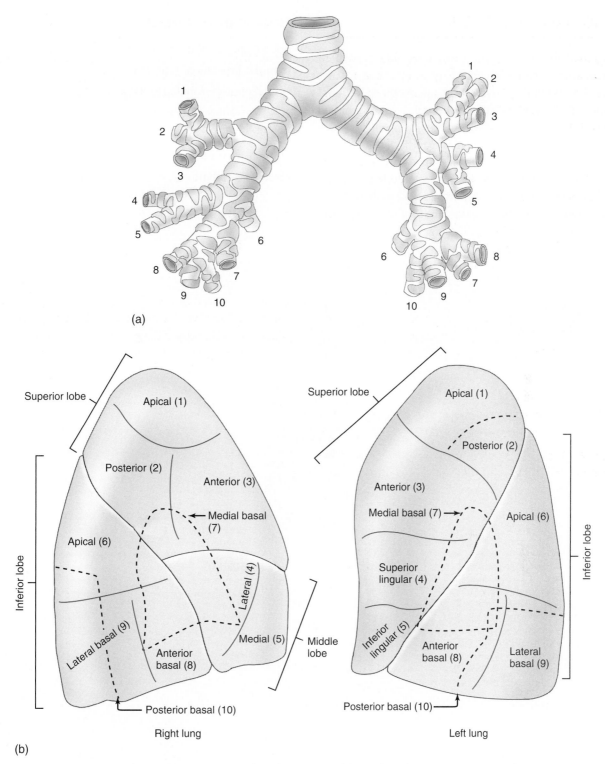

(a)

(b)

FIG. 18.4 Respiratory tree, anterior view (a), and bronchopulmonary segments (b).

GAS EXCHANGE SURFACES

The final divisions of the respiratory tree are called **respiratory bronchioles** and these give rise to tiny **alveolar ducts** that have a thin squamous endothelial lining. The alveolar ducts have lateral pockets called **alveoli**. It is at the alveoli that gas exchange takes place with the blood in the pulmonary capillaries. There are about 300 million alveoli in the lungs with a total surface area of 700 square feet. The endothelial membranes of the lungs and capillaries, along with the basement membranes, have a thickness of only 0.5 μm, allowing for a very rapid exchange of gases.

Respiratory Lobule

Locate and *label* the following in Figure 18.5.

terminal bronchiole (BRONG-ke-ol)

 respiratory lobule (LOB-ul)

  alveolus (al-VE-o-lus)

pulmonary arteriole

pulmonary venule

alveolar capillaries

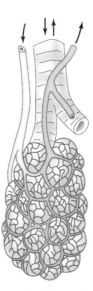

FIG. 18.5 Respiratory lobule.

SHORT ANSWER

1. What organs are part of the upper respiratory tract? The lower respiratory tract?

2. *Describe* the location of the meatuses. How do they relate to the paranasal sinuses?

3. *List* the layers through which a needle passes between the epidermis and the airspace inside an alveolus.

4. *Describe* the airway supplying the following:

   right lung

   lung lobe

   bronchopulmonary segment

respiratory lobule

alveolus

5. *Trace* inhaled air from the external nare to the alveolus by listing each structure in the pathway.

MATCHING

Match the descriptions in column A with the terms in column B.

<table>
<tr><td align="center">A</td><td align="center">B</td></tr>
<tr><td>_____ 1. roof of the oral cavity reinforced with bone</td><td>a. adenoid</td></tr>
<tr><td>_____ 2. nasal bones that increase surface area</td><td>b. alveolus</td></tr>
<tr><td>_____ 3. region of pharynx posterior to the nasal cavity</td><td>c. bronchi</td></tr>
<tr><td>_____ 4. opens to middle meatus segments</td><td>d. bronchopulmonary</td></tr>
<tr><td>_____ 5. cartilage that prevents food from entering larynx</td><td>e. conchae</td></tr>
<tr><td>_____ 6. lymph tissue in nasopharynx</td><td>f. epiglottis</td></tr>
<tr><td>_____ 7. divisions of the trachea</td><td>g. frontal sinus</td></tr>
<tr><td>_____ 8. last bronchial division</td><td>h. hard palate</td></tr>
<tr><td>_____ 9. site of gas exchange with the blood</td><td>i. nasopharynx</td></tr>
<tr><td>_____ 10. membranes covering the lung</td><td>j. pleura</td></tr>
<tr><td>_____ 11. sound-producing membrane of the larynx</td><td>k. respiratory bronchiole</td></tr>
<tr><td>_____ 12. produces pulmonary surfactant</td><td>l. type II (alveolar) cells</td></tr>
<tr><td>_____ 13. supplied by tertiary bronchioles</td><td>m. vocal fold</td></tr>
</table>

MULTIPLE CHOICE

_____ 1. The Adam's apple is a common term for the
   a. pharynx.
   b. epiglottis.
   c. cricoid cartilage.
   d. thyroid cartilage.

_____ 2. Surfactant is produced by
   a. alveolar basement membrane.
   b. type I (pulmonary epithelial) cells.
   c. type II (septal) cells.
   d. blood cells.

_____ 3. Pharyngeal tonsils are located in the
   a. nasopharynx.
   b. oropharynx.
   c. laryngopharynx.
   d. larynx.

_____ 4. The laryngeal cartilage involved in sound production is the
   a. thyroid.
   b. epiglottis.
   c. cricoid.
   d. arytenoid.

_____ 5. Posteriorly, the nasal cavity communicates with the pharynx through the
   a. vestibule.
   b. auditory tube.
   c. internal nares.
   d. superior meatus.

Contents

Objectives

1. List the components and describe the anatomy of the oral cavity. Include the salivary glands and tongue.

2. List the kinds and numbers of teeth.

3. Describe the anatomy of a typical tooth.

4. Describe the anatomy of the stomach. Describe the histology and secretions of the wall.

5. Describe the anatomy of the duodenum. List the hormones produced by the duodenal wall and their effects on the target organs.

6. List the organs associated with the duodenum.

7. Describe the anatomy and functions of the small intestine, large intestine, rectum, colon, and anus.

8. Trace the pathway through the digestive system of an ingested saltine cracker. List the structures involved in digestion and absorption. Indicate the enzymes involved.

FUNCTIONS OF THE DIGESTIVE SYSTEM

The digestive system is a "disassembly" system. It functions to reduce ingested food from large, complex molecules to a size that can be absorbed through the digestive endothelium into the blood or lacteals. Digestion takes two forms, chemical and mechanical. **Chemical**

digestion is a process in which carbohydrates, proteins, and fats are changed in chemical structure. **Mechanical digestion** reduces food particles in size by mechanical means such as mastication.

The **gastrointestinal (GI) tract** begins in the oral cavity and includes the esophagus, stomach, small intestine, large intestine, rectum, and anus. A number of accessory structures such as glands, gallbladder, tongue, and teeth are included. The endothelium of the GI tract is called the **mucosa**. It is mostly of simple arrangement except in the mouth and esophagus where it is stratified. Much of the mucosa is secretory as well as absorptive. It produces mucus for protection of the endothelium and, in the duodenum and stomach, it secretes digestive enzymes.

THE ORAL CAVITY

Locate and *label* the following in Figure 19.1.

lingual (LING-gwal) and palatal (PAL-ah-tal)

    incisors (in-SI-zers)

      central

      lateral

    canines (KA-nin) or cuspids (KUS-pids)

    premolars (pre-MO-lar) or bicuspids

      1st

      2nd

    molars

      1st

      2nd

      3rd (wisdom teeth)

      gingivae (jin-JI-ve)

    tongue

      papillae

      circumvallate (ser-kum-VAL-at)

      filiform (FIL-ih-form)

      fungiform (FUN-jih-form)

    hard palate (PAL-at)

    soft palate

      uvula (U-vyoo-lah)

    palatine tonsil

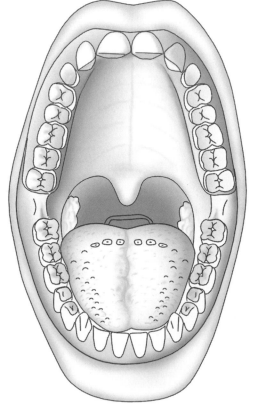

FIG. 19.1 Oral cavity, anterior view.

Cuspid Tooth, Sagittal Section

Locate and *label* the following in Figure 19.2.

crown

    enamel (en-AM-el)

    dentin (DEN-tin)

neck

    pulp cavity

root

    apical (AP-eh-kal) foramen

Refer to your textbook or other source and *describe* the location of the following salivary glands:

parotid (pah-ROT-id) _____

_____

submandibular _____

_____

sublingual _____

_____

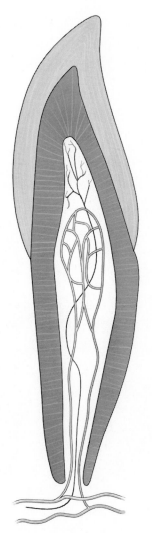

FIG. 19.2 Cuspid tooth, sagittal section.

Figure 19.3 illustrates components of the digestive and skeletal systems. *Locate* the various structures using skeletal landmarks as a reference. *Determine* the position of the organs in your own body by *palpating* the skeletal landmark through the skin.

STOMACH AND SMALL INTESTINE

Stomach and Small Intestine, Anterior View

Locate and *label* the following in Figure 19.3.

esophagus (e̱-SOF-ah-gus)

    gastroesophageal (gas-tro̱-e̱-sof-ah-JE̱-al) or cardiac sphincter (SFINK-ter)

stomach

    cardia

    fundus (FUN-dus)

    body

    pylorus (pi̱-LO̱-rus)

    lesser curvature

    greater curvature

    rugae (ROO-je̱)

    gastric glands   ⎫ (not illustrated;
        chief cells   ⎬ describe the location
        parietal cells ⎭ and function)

    mucous cells

small intestine

    pyloric sphincter

    duodenum (doo-o̱-DE̱-num)

    jejunum (jeh-JOO-num)

    ileum (IL-e̱-um)

    ileocecal (IL-e̱-o̱-se̱-kal) sphincter

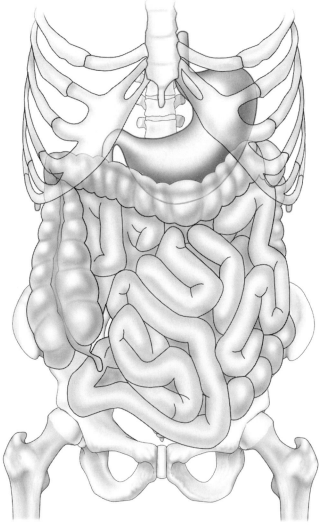

FIG. 19.3 Stomach and small intestine, anterior view.

Small Intestine, Transverse Section

Locate and *label* the following in Figure 19.4.

tunics (TOO-niks)

  mucosa (myoo-K<u>O</u>-sah)

    villi (VIL-<u>i</u>)

    lacteal (LAK-t<u>e</u>-al)

    villiary capillaries

  submucosa

  muscularis (mus-kyoo-L<u>A</u>-ris)

    circular muscle

    longitudinal muscle

  serosa (seh-R<u>O</u>-sah)

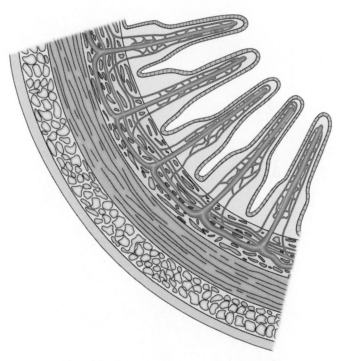

FIG. 19.4 Small intestine, transverse section.

COLON AND ACCESSORY GLANDS

Large Intestine and Accessory Glands, Anterior View

Locate and *label* the following in Figure 19.5.

liver

pancreas

large intestine

    taenia coli (T<u>E</u>-n<u>e</u>-ah K<u>O</u>-li)

    haustra (HAWS-trah)

    cecum (S<u>E</u>-kum)

        vermiform appendix (VER-mih-f<u>o</u>rm ah-PEN-diks)

    colon (K<u>O</u>-lon)

        ascending

        transverse

        descending

        sigmoid (SIG-moyd)

    rectum (REK-tum)

    anus (<u>A</u>-nus)

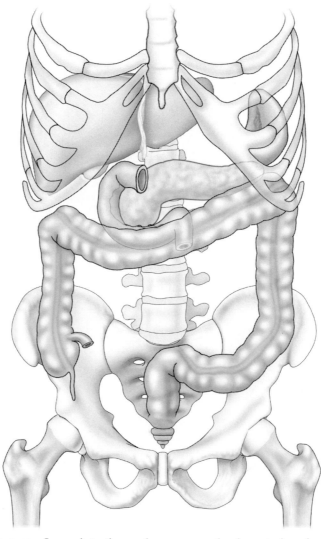

FIG. 19.5 Large intestine and accessory glands, anterior view.

Liver, Gallbladder, Pancreas, and Duodenum, Anterior View

Locate and *label* the following in Figure 19.6.

duodenum
pancreas
    pancreatic duct
liver
    right lobe
    left lobe
    falciform (FAL-sih-f<u>o</u>rm) ligament
    gallbladder
        common hepatic duct
        cystic (SIS-tik) duct
        common bile (b<u>i</u>l) duct

The liver produces bile from the breakdown of red blood cells. Bile is directed to the duodenum where it functions to emulsify fats. When a meal is not undergoing digestion, bile is stored and concentrated by the gall-bladder. When bile is required, the **sphincter of Oddi** (<u>O</u>D-<u>e</u>), near the entrance to the duodenum, opens under the influence of **cholecystokinin** (k<u>o</u>-l<u>e</u>-sis-t<u>o</u>-K<u>I</u>N-in), a duodenal wall hormone, and bile flows to the duodenum.

CAT DISSECTION

If you have not yet opened the thoracic and abdominal cavities of the cat, do so now. *Refer* to the section on opening the body cavity in Chapter 13.

You need not identify the teeth in the cat; use a human skull instead. On the cat, *find* the **tongue** and on it, the **circumvallate papillae.** From here *look laterally* to the **palatine tonsils.** You should also *examine* the top of the oral cavity to find the **hard palate** and **soft palate.** Behind the oral cavity is the **oropharynx.**

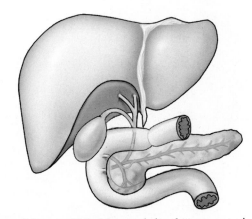

FIG. 19.6 Liver, gallbladder, and duodenum, anterior view.

Refer to Figure 19.7 and CDA-18 and *locate* the **esophagus** deep to the trachea and *follow* it through the diaphragm to the stomach. As you begin to identify organs in the abdomen, you may find the **greater omentum**, a broad, fat-laden membrane covering the region. *Note* the position of the **gastroesophageal sphincter** at the point where the esophagus joins the stomach. Also *find* the **pyloric sphincter**, a similar sphincter at the inferior end of the stomach. *Make* an incision down the length of the stomach and open it to find folds in the endothelium called **rugae**. The section of the small intestine leaving the stomach then looping to the left and below it is the **duodenum**. This section is followed by the **jejunum** and then the **ileum** which terminates at the **large intestine** deep on the right side of the animal. *Find* the **ileocolic sphincter** at this junction. *Note* the **cecum**, a blind pouch immediately beyond the sphincter. The large intestine runs craniad and then caudad to join the **rectum**, an enlargement just proximal to the **anus**. *Examine* the **liver** and *note* that it is composed of several **lobes**. *Retract* the lobes and find the **gallbladder**. See if you can find the **common bile duct** (green in color) running to the duodenum. *Use* your scissors and *make* a clean cut through one of the lobes of the liver. If your cat is triple injected, you will see the **hepatic veins** and **central veins** as blue and the **sinusoids** and **hepatic portal vessels** as yellow.

Now *look* caudad to the stomach and note the pinkish **pancreas**. Take some time to *trace* the extent of this gland, as it is very diffuse.

Locate the small, round **intestinal lymph nodes** distributed throughout the **intestinal mesenteries**. Finally, *find* the three-inch long, half-inch wide **spleen** on the left side of the stomach.

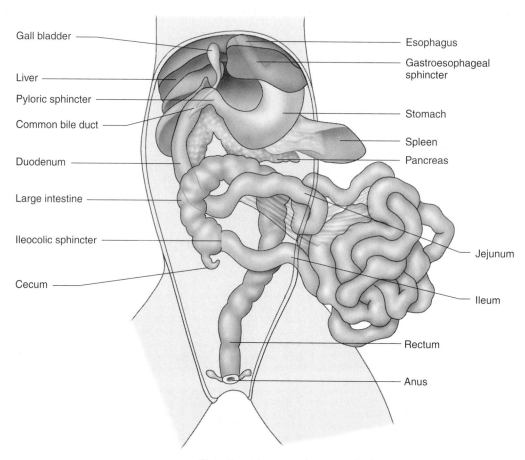

FIG. 19.7 Cat, digestive system, ventral view.

SHORT ANSWER

1. *List* the components of the digestive system. Indicate the digestive processes that occur at each.

2. How do mechanical and chemical digestion differ?

3. *Describe* the differences in structure of the wall of the esophagus, stomach, small intestine, and large intestine.

4. *List* the products produced by the pancreas, stomach wall, and duodenal wall.

5. *Sketch* the connections between the liver, gallbladder, pancreas, and duodenum.

MATCHING

Match the descriptions in column A with the terms in column B.

A

_____ 1. salivary gland medial to angle of jaw

_____ 2. sphincter at top of stomach

_____ 3. increases the surface area of small intestine

_____ 4. secretory portion of small intestine

_____ 5. secretory portion of stomach

_____ 6. sphincter at distal end of small intestine

_____ 7. controls bile secretion

_____ 8. carries bile to the duodenum

_____ 9. exocrine and endocrine gland

_____ 10. emulsifies fats

_____ 11. water absorption

_____ 12. longest section of large intestine

_____ 13. decreases gastric activity

_____ 14. stimulates pancreatic activity

_____ 15. blind sac at origin of colon

B

a. bile

b. cecum

c. cholecystokinin

d. colon

e. common bile duct

f. duodenum

g. enterogastrone

h. fundus

i. gastroesophageal

j. ileocecal

k. submandibular

l. microvilli

m. pancreas

n. secretin

o. transverse colon

MULTIPLE CHOICE

_____ 1. An incision into the small intestine would cut which layer first?

    a. muscularis
    b. mucosa
    c. submucosa
    d. serosa

_____ 2. The colon appears subdivided into pouches called

    a. taenia coli.
    b. villi.
    c. rugae.
    d. haustra.

_____ 3. The salivary glands located anterior and inferior to the external auditory meatus are the

    a. submandibular.
    b. parotid.
    c. sublingual.
    d. buccal.

_____ 4. The stomach contacts the left side of the body here.

    a. greater curvature
    b. lesser curvature
    c. cardiac portion
    d. pyloric portion

_____ 5. The terminal portion of the small intestine is the

    a. duodenum.
    b. ileum.
    c. jejunum.
    d. anus.

Contents

Objectives

1. Describe the location and gross anatomy of the kidney.
2. Trace the flow of blood through the kidney and name the vessels.
3. Describe the anatomy and function of each region of the nephron.
4. Describe the location and gross anatomy of the ureters, urinary bladder, and urethra.
5. Describe the gross anatomy, location, and function of each structure in the male and female reproductive systems. Include external genitalia and accessory glands.
6. Define and describe meiosis.
7. Describe oogenesis and spermatogenesis.
8. Describe the process of fertilization and early development.

FUNCTIONS OF THE UROGENITAL SYSTEM

The **urogenital system** includes both **reproductive** and **excretory** functions. In the male some structures share both functions.

The kidney performs the following functions:

1. Removes wastes from the blood.
2. Regulates blood pH.
3. Regulates blood pressure.
4. Controls the concentrations of blood solutes.

The kidneys are located posterior to the abdominal cavity in a **retroperitoneal** position. Figure 20.1 shows the bony landmarks you should use to locate the kidneys.

Urine is conducted by the ureters to the urinary bladder where it is emptied to the outside through the urethra.

EXCRETORY ORGANS

Excretory Organs, Posterior View

Locate and *label* the following in Figure 20.1.

kidney

ureter (yoo-R<u>E</u>-ter)

urinary bladder

urethra (yoo-R<u>E</u>-thrah)

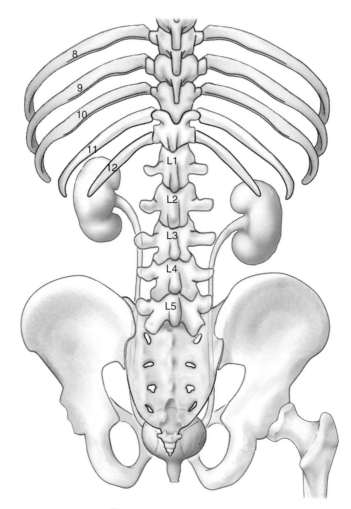

FIG. 20.1 Excretory organs, posterior view.

Kidney, Frontal Section

Locate and *label* the following in Figure 20.2.

renal cortex (RE-nal KOR-teks)

renal medulla (me-DUL-ah)

    renal pyramid

    renal papilla

renal pelvis

    renal calyces (KAL-ih-sez)

        primary calyx

        secondary calyx

ureter

renal artery

renal vein

interlobar (in-ter-LO-bar) artery and vein

arcuate (AR-kyoo-at) artery and vein

interlobular (in-ter-LOB-yoo-lar) artery and vein

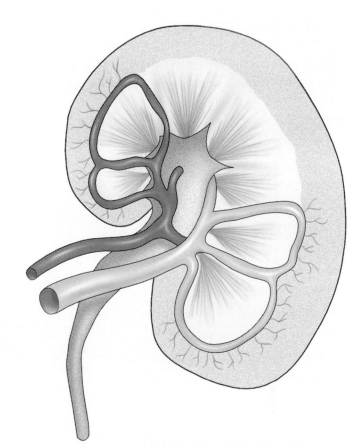

FIG. 20.2 Kidney, frontal section.

Nephron, Schematic View

Locate and *label* the following in Figure 20.3.

glomerular (glo-MER-yoo-lar) or Bowman's capsule

proximal convoluted (KON-vo-loot-ed) tubule

loop of Henle (HEN-le)

distal convoluted tubule

afferent arteriole

glomerulus (glo-MER-yoo-lus)

efferent arteriole

collecting duct

juxtaglomerular (juks-tah-glo-MER-yoo-lar) apparatus

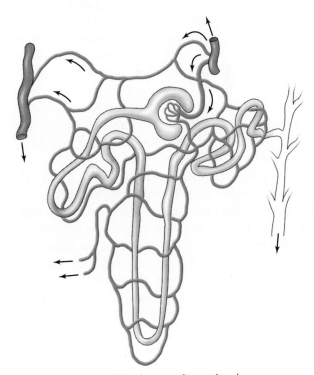

FIG. 20.3 Nephron, schematic view.

KIDNEY DISSECTION

Obtain a representative mammalian kidney from your instructor, *rinse* it, and *section* it in the frontal plane. While *referring* to Figure 20.2, *locate* the following:

renal cortex

renal medulla

    renal pyramid

    renal papilla

renal pelvis

    renal calyces

ureter

renal artery

    interlobar artery

      arcuate artery

renal vein

    interlobar vein

      arcuate vein

REPRODUCTIVE ORGANS

The function of the reproductive system is the continuation of the species. In both sexes the **gonads** produce **haploid cells** by the process of **meiosis**. These cells contain half of the adult chromosome complement of 46 (thus, 23). When the male gamete, the **sperm**, unites with female gamete, the **ovum**, in **fertilization**, the **diploid** number of 46 is restored. Fertilization occurs in the upper one-third of the **uterine tube**. This first diploid cell of the new individual is called the **zygote**. The new cell cleaves until a solid ball of cells, the **morula**, is formed. The morula is the same size as the original zygote. While the cell number is increasing during this time, the size does not. The process by which cell number increases but cell size does not is called **cleavage**. As the morula drifts down the uterine tube to the uterus it is converted to a hollow ball of cells, the **blastocyst**, which will implant in the mother's **endometrium**. This process takes about one week. The association with the endometrium will allow the blastocyst to obtain nutrients from the mother's blood and to get rid of waste products in the same manner.

During the first two months, the new organism is called an **embryo**. After this time and until birth it is called a **fetus**. Shortly after implantation the embryo de-velops three primary germ layers and is then called a **gastrula**. These germ layers are the **ectoderm, endoderm**, and **mesoderm**. At this time, a number of cell migrations and differentiations occur. The **amnionic** cavity forms and will eventually contain **amnionic fluid** that is released as the membrane ruptures just prior to birth as the "water breaking." The **yolk sac** is evident but nonfunctional in the human. The **chorion** enlarges and will form the **placenta** through which the embryo and then the fetus will exchange nutrients and waste products with the mother's blood. At this time the **allantoic membrane** is also forming and will ultimately become the **umbilical cord** connecting the placenta to the fetus. The forty weeks from fertilization to **parturition** is called **gestation**. During this time all organs and organ systems are formed. Take time to *read* your textbook and *see* what changes occur during each month of development. When gestation is shortened, some systems may not be fully formed or be ready to assume a function outside the uterus. Complications often accompany premature births. A good diet during pregnancy accompanied by mild exercise, regular medical checkups, and prenatal classes are all that is generally required to produce a healthy baby. Alcohol, tobacco products, and other drugs, including vitamin and mineral supplements and other nonprescription drugs not prescribed by a physician are to be avoided completely during all stages of pregnancy.

HUMAN DEVELOPMENT

Locate and *label* the following in Figure 20.4.

zygote (ZI-got)

two-cell stage

morula (MOR-u-lah)

blastocyst (BLAS-to-sist)

fourteen-day embryo

    amnion (AM-ne-on)

    amnionic cavity

    embryonic disk

      ectoderm (EK-to-derm)

      mesoderm (MES-o-derm)

      endoderm (EN-do-derm)

    yolk sac

    chorion (KO-re-on)

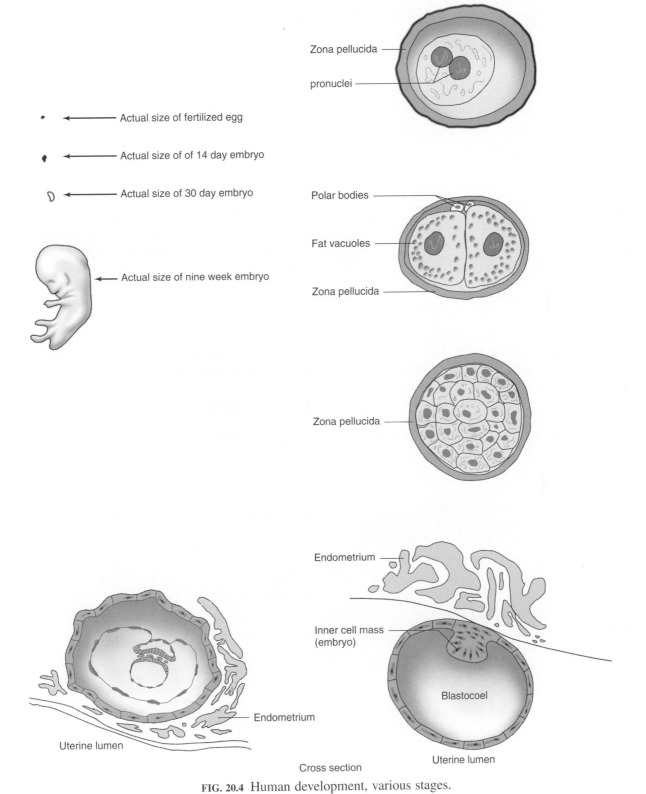

Zona pellucida

pronuclei

Actual size of fertilized egg

Actual size of of 14 day embryo

Actual size of 30 day embryo

Actual size of nine week embryo

Polar bodies

Fat vacuoles

Zona pellucida

Zona pellucida

Endometrium

Inner cell mass
(embryo)

Blastocoel

Endometrium

Uterine lumen

Cross section

Uterine lumen

FIG. 20.4 Human development, various stages.

Female Urogenital System, Midsagittal Section

Locate and *label* the following in Figure 20.5.

sacrum

coccyx

pubis

rectum

ureter

urinary bladder

    urethra

        urethral orifice

anus

    anal sphincter

ovary

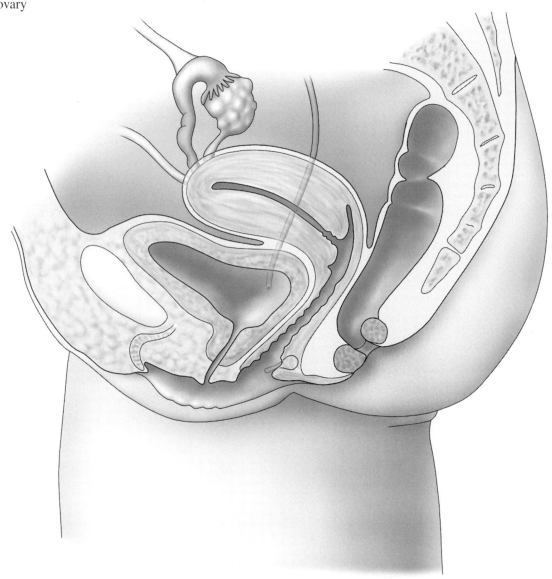

FIG. 20.5 Female urogenital system, midsagittal section.

uterine (YOO-ter-in) (fallopian) tube

ovarian (o-VA-re-an) ligament

uterus (YOO-ter-us)

vagina (vah-JI-nah)

    fornix (FOR-niks)

    orifice

        hymen (HI-men)

        vestibular (ves-TIB-yoo-lar) gland

external genitalia

    labia (LA-be-ah) majora

    labia minora

    vestibule (VES-tih-bul)

    clitoris (KLI-to-ris)

    prepuce (PRE-pus)

    mons pubis (mahnz PYOO-bis)

Uterus, Frontal Section

Locate and *label* the following in Figure 20.6.

ovary

uterine or fallopian tube

    infundibulum (in-fun-DIB-yoo-lum)

        fimbriae (FIM-bre-e)

uterus

    fundus (FUN-dus)

    body or corpus (KOR-pus)

    cervix (SER-viks)

    perimetrium (per-ih-ME-tre-um)

    myometrium (mi-o-ME-tre-um)

    endometrium (en-do-ME-tre-um)

    uterine cavity (lumen)

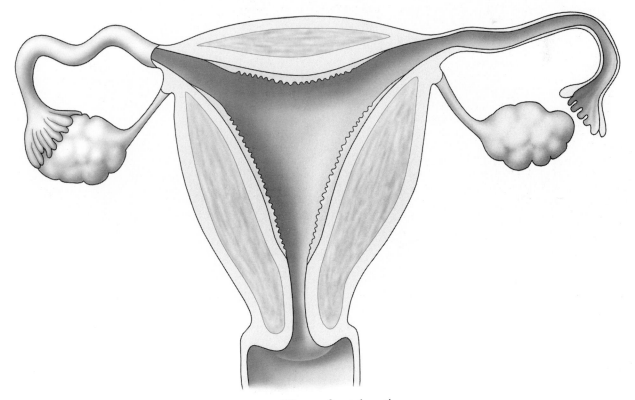

FIG. 20.6 Uterus, frontal section.

Vulva, Inferior View

Locate and *label* the following in Figure 20.7.

prepuce of clitoris

clitoris

vestibule

labia minora

vaginal orifice

urethral orifice

labia majora

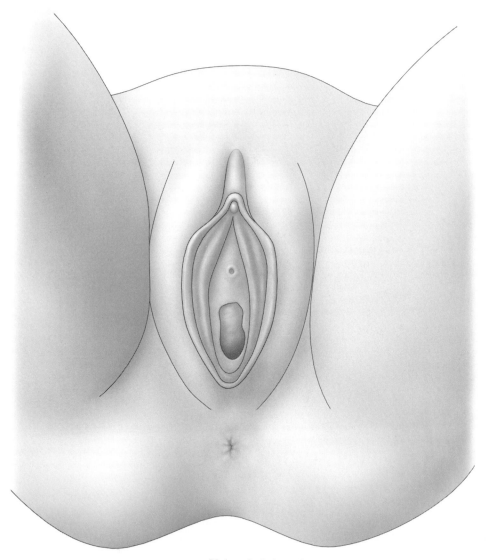

FIG. 20.7 Vulva, inferior view.

Oogenesis

Figure 20.8 illustrates the process of oogenesis (o-o-JEN-eh-sis). A diploid primary oocyte (O-o-sit) is converted to a haploid ovum (O-vum) by the process of meiosis (mi-O-sis). When a female is born, she has already produced all of her primary oocytes for life. This occurs during the third month of gestation. Meiosis continues at puberty and produces one fertile ovum per month during the reproductive life of the individual.

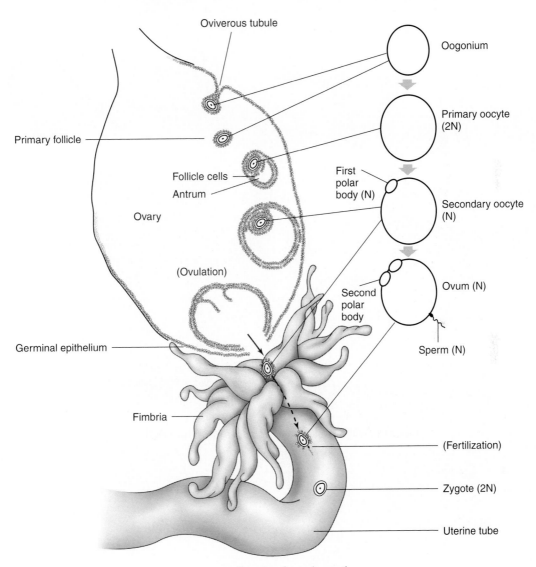

FIG. 20.8 Oogenesis, schematic.

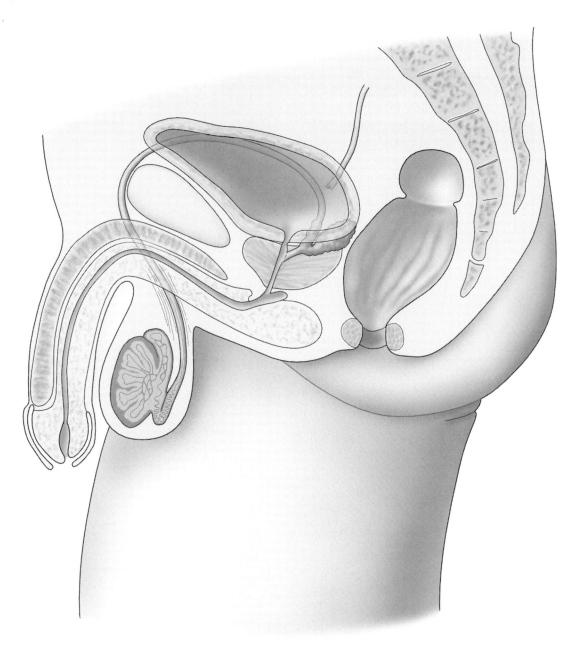

FIG. 20.9 Male urogenital system, midsagittal section.

MALE UROGENITAL SYSTEM

Male Urogenital System, Midsagittal Section

Locate and *label* the following in Figure 20.9.

pubis

sacrum

coccyx

rectum

anus

    anal sphincter

ureter

urinary bladder

    urethra

scrotum (SKRO-tum)

    cremaster (kre-MAS-ter) muscle

testis (TES-tis)

    seminiferous (seh-mih-NIF-er-us) tubule

epididymis (ep-ih-DID-ih-mis)

ductus (VAS) deferens (DUK-tus DEF-er-enz)

seminal vesicle (SEM-ih-nal VES-ih-kl)

ejaculatory (e-JAK-yoo-lah-to-re) duct

prostate (PROS-tat) gland

bulbourethral (bul-bo-yoo-RE-thral) gland

penis (PE-nis)

    glans (glanz) penis

    urethral orifice

    prepuce or foreskin

    corpora cavernosa (KOR-po-rah kav-er-NO-sah)

    corpus spongiosum (KOR-pus spon-je-O-sum)

Penis, Transverse Section

Locate and *label* the following in Figure 20.10.

dorsal blood vessels

corpora cavernosa

corpus spongiosum

urethra

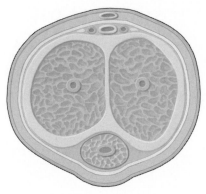

FIG. 20.10 Penis, transverse section.

Spermatogenesis

Unlike oogenesis in the female, **spermatogenesis** (sper-ma-to-JEN-eh-sis) in the male does not begin until puberty. From each diploid **primary spermatocyte** (sperm-MA-to-sit), four sperm cells are formed in a process that continues throughout life. In the female, around 450 mature ova are produced during a reproductive lifetime. In the male, 350 million sperm are produced per day. Figure 20.11 illustrates the process of spermatogenesis.

Sperm Cell

Locate and *label* the following in Figure 20.12.

head

    acrosome (AK-ro-som)

    nucleus

midpiece

    mitochondria

tail or flagellum

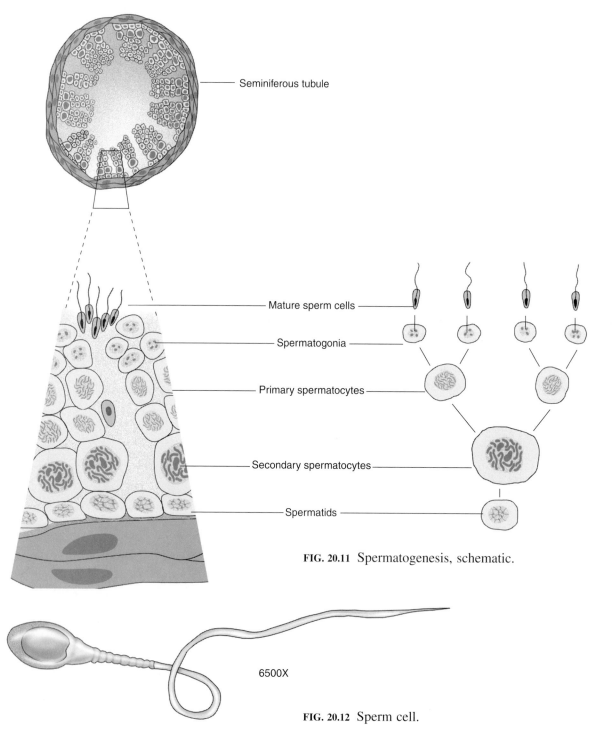

Seminiferous tubule

Mature sperm cells

Spermatogonia

Primary spermatocytes

Secondary spermatocytes

Spermatids

FIG. 20.11 Spermatogenesis, schematic.

6500X

FIG. 20.12 Sperm cell.

CAT DISSECTION

Cat Urogenital System, Anterior Views

Find a cat of the opposite sex from your specimen at another lab table so that you are able to observe the urogenital system in both sexes.

Locate the following features on the cat shown in Figure 20.13(a) and (b). Refer to CDA-19 and CDA-20.

kidney

ureter

urethra

urinary bladder

male cat

    scrotum

    testis

    epididymis

    penis

female cat

    ovary

    uterine tube

    uterus

    vagina

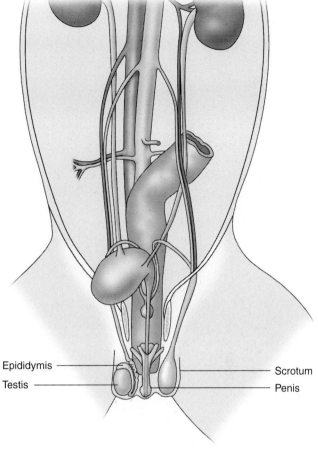

(a)

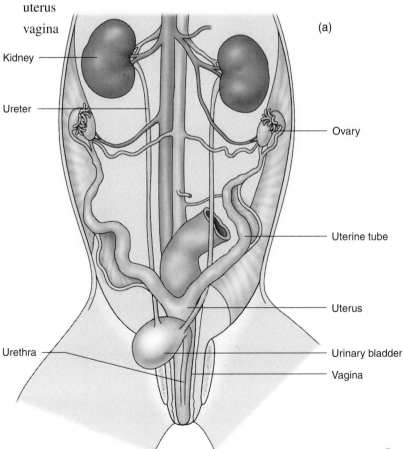

(b)

FIG. 20.13 Cat urogenital system male (a) and female (b).

SHORT ANSWER

1. *Place* the following in the correct order from superficial to deep.

   a. epidermis
   b. lumen of the uterus
   c. endometrium
   d. myometrium
   e. dermis
   f. perimetrium
   g. parietal peritoneum
   h. visceral peritoneum
   i. body cavity
   j. subcutaneous layer
   k. body wall musculature

2. *Place* the following in the correct order beginning where filtrates leave the blood.

   a. ureter
   b. collecting duct
   c. pelvis
   d. glomerular capsule
   e. renal loop
   f. distal convoluted tubule
   g. primary calyx
   h. proximal convoluted tubule
   i. urethra
   j. urinary bladder
   k. renal papillum
   l. secondary calyx

3. *Draw* and *label* the parts of a sperm cell.

4. *Trace* a sperm cell from the seminiferous tubules to the urethra. *Indicate* the location of the accessory glands making contributions to the semen.

MATCHING

Match the descriptions in column A with the terms in column B. Responses I and J may be used more than once.

A

_____ 1. site of reabsorption of materials to the blood

_____ 2. portion of nephron where greatest reabsorption occurs

_____ 3. blood supply leading to the glomerulus

_____ 4. blood vessel draining to interlobular vein

_____ 5. capillary network in renal capsule

_____ 6. portion of tubule receiving material from glomerulus

_____ 7. tubule draining to a collecting duct

_____ 8. found in 15% of all human nephrons

_____ 9. receives outflow of several nephrons

_____ 10. found in the renal cortex

B

a. afferent arteriole

b. collecting duct

c. distal convoluted tubule

d. glomerulus

e. loop of Henle

f. peritubular capillaries, vasa recta

g. proximal convoluted tubule

h. renal capsule

i. more than one of the above

j. none of the above

Match the term in column A with the homologous structures in column B.

A

_____ 1. penis

_____ 2. scrotum

_____ 3. penile urethra

_____ 4. bulb of the penis

_____ 5. prostate gland

_____ 6. testis

_____ 7. bulbourethral gland

_____ 8. glans penis

B

a. bulb of the vestibule

b. clitoris

c. glans clitoris

d. greater vestibular glands

e. labia majora

f. labia minora

g. ovary

h. paraurethral glands

MULTIPLE CHOICE

_____ 1. Which best defines *meiosis*?

   a. production of haploid gametes
   b. production of primary germ cells
   c. tissue repair
   d. formation of the endometrium

_____ 2. Which of the following is altered by a vasectomy?

   a. testis
   b. ejaculatory duct
   c. urethra
   d. ductus deferens

_____ 3. The structure surgically removed by circumcision is the

   a. bulb.
   b. prepuce.
   c. glans.
   d. testis.

_____ 4. Ova and sperm differ from all other body cells in that they

   a. have the haploid chromosome number.
   b. undergo mitotic division.
   c. cannot develop or mature.
   d. contain all cell organelles except mitochondria.

_____ 5. Fertilization normally occurs in the

   a. uterine (Fallopian) tubes.
   b. uterus.
   c. ovaries.
   d. vagina.

_____ 6. The advantage of sexual reproduction is that

   a. it produces variation within the species.
   b. it ensures that all members of the species will be genetically identical.
   c. it produces a zygote from two diploid cells.
   d. the chromosome number is doubled in each generation.

_____ 7. During a menstrual cycle, a sudden drop in plasma estrogen levels indicates

   a. pregnancy.
   b. menstruation will soon begin.
   c. ovulation.
   d. the end of menstruation.

_____ 8. Which of the following maintains the corpus luteum during pregnancy?

   a. LH
   b. FSH
   c. HCG
   d. estrogen

_____ 9. Which of the following initiates follicular development?

   a. LH
   b. FSH
   c. HCG
   d. estrogen

_____ 10. Which of the following is responsible (indirectly) for increasing blood pressure?

   a. FSH
   b. LH
   c. enterogastrone
   d. renin

21 Topography

Contents

21

Objectives

1. Describe surface anatomy of each body region.
2. Correlate surface features with deeper anatomy.
3. Understand how the regions of the body relate.
4. Review the contents of the chapters that play a part in surface anatomy.
5. Provide information necessary to properly dissect the human cadaver.

INTRODUCTION TO TOPOGRAPHY

Topography (to-POG-rah-fe) or surface anatomy is the study of the form and markings appearing on the surface of the body. Throughout this manual, you have already located many topographical features by palpating those items marked with an asterisk (*). Additional terms that will serve to complement your already extensive understanding of anatomy follow below. If a term has appeared in a previous chapter, that chapter is referenced.

Use the appropriate chapters in your text or charts of surface anatomy as you complete the following exercises.

Head, Lateral View

Locate and *label* the following in Figure 21.1.

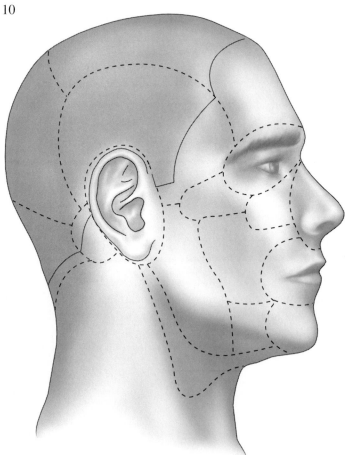

FIG. 21.1 Head, lateral view.

Face, Anterior View

Locate and *label* the following in Figure 21.2.

orbital region	Ch. 7
pupil	Ch. 16
iris	Ch. 16
sclera	Ch. 16
palpebra or eyelid	Ch. 16
medial commissure	
lateral commissure	
lacrimal caruncle (KAR-ung-kl)	
eyelash	Ch. 16
supercilia (syoo-per-SIL-e̱-ah) or eyebrows	Ch. 16
nose and lips	
root	
apex (A̱-peks)	
dorsum nasi	
ala (A̱-lah)	
external nars	Ch. 18
bridge	
philtrum (FIL-trum)	
lips	

FIG. 21.2 Face, anterior view.

Pinna, Lateral View

Locate and *label* the following in Figure 21.3.

pinna (PIN-na) or auricle	Ch. 16
tragus (TR<u>A</u>-gus)	
antitragus	
concha (KONG-ka)	
helix (H<u>E</u>-liks)	
external auditory meatus	Ch. 6
lobule	

NECK

Neck, Anterior View

Locate and *label* the following in Figure 21.4.

trapezius muscle	Ch. 10
cricoid cartilage of larynx	Ch. 18
thyroid cartilage of larynx	Ch. 18
hyoid	Ch. 6
trachea	Ch. 18
sternocleidomastoid muscle	Ch. 10
external jugular vein	Ch. 12

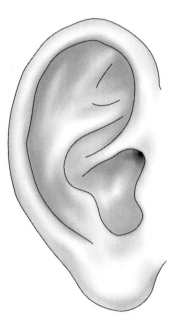

FIG. 21.3 Pinna, lateral view.

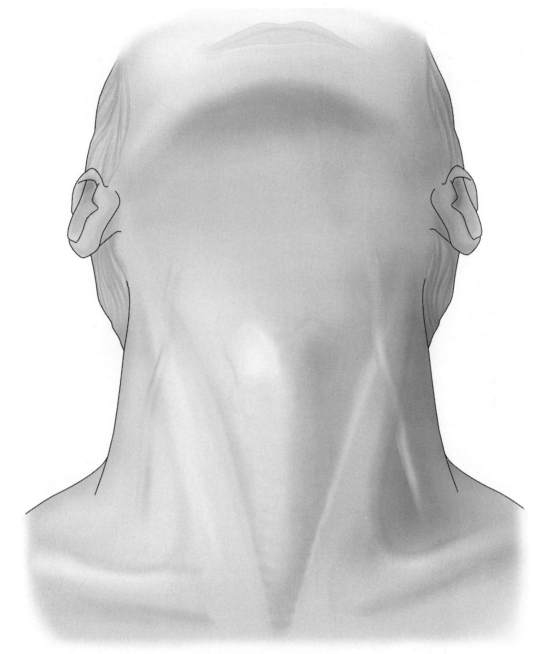

FIG. 21.4 Neck, anterior view.

TRUNK

Trunk, Anterior View

Locate and *label* the following in Figure 21.5.

chest

clavicle	Ch. 7
jugular notch of sternum	Ch. 6
manubrium of sternum	Ch. 6
serratus anterior muscle	Ch. 10
nipple	
anterior axillary fold	
pectoralis major muscle	Ch. 10
ribs	Ch. 6
abdomen	Ch. 1
umbilicus (um-BIL-i-kus)	Ch. 4
rectus abdominis muscle	Ch. 10
external oblique muscle	Ch. 10
linea alba (LIN-e̠-ah AL-bah)	Ch. 10
pubic symphysis	Ch. 7
anterior superior iliac spine	Ch. 7

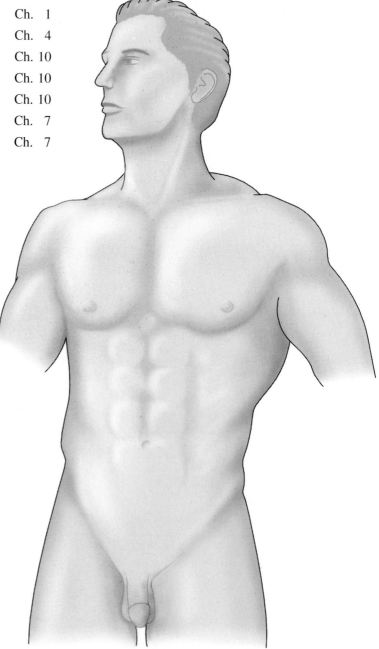

FIG. 21.5 Trunk, anterior view.

Trunk, Posterior View

Locate and *label* the following in Figure 21.6.

back	Ch. 1
vertebral spines	Ch. 6
scapula	Ch. 7
latissimus dorsi muscle	Ch. 10
erector spinae muscle	Ch. 10
trapezius muscle	Ch. 10
buttocks and pelvis	
iliac crest	Ch. 6
posterior superior iliac spine	Ch. 6
sacrum	Ch. 6
gluteus medius muscle	Ch. 10
gluteus maximus muscle	Ch. 10
gluteal cleft	
gluteal fold	
greater trochanter	Ch. 6

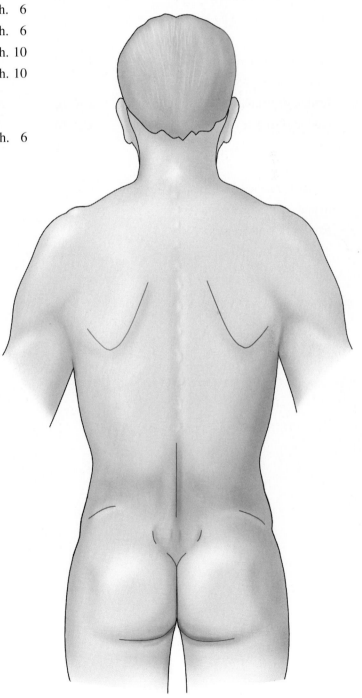

FIG. 21.6 Trunk, posterior view.

SUPERIOR APPENDAGE

Superior Appendage, Anterior View

Locate and *label* the following in Figure 21.7(a).

brachium	Ch. 1
acromion	Ch. 7
deltoid muscle	Ch. 10
biceps brachii muscle	Ch. 10
bicipital aponeurosis (bi-SIP-ih-tal ap-o-noo-RO-sis)	
brachioradialis muscle	Ch. 10
cubital fossa (KYOO-bih-tl)	
antebrachium	Ch. 1
median cubital vein	Ch. 12
wrist creases	
tendon of palmaris longus muscle	Ch. 10
tendon of flexor carpi radialis muscle	Ch. 10
manus	Ch. 1
pisiform	Ch. 7
thenar eminence (THE-nar EM-ih-nens)	
hypothenar eminence	
palmar flexion creases	
digital flexion creases	

Superior Appendage, Posterior View

Locate and *label* the following in Figure 21.7(b).

brachium	Ch. 1
triceps brachii muscle	Ch. 7
medial and lateral epicondyle	Ch. 7
antebrachium	Ch. 1
olecranon process	Ch. 7
styloid process of radius	Ch. 7
manus	Ch. 1
tendon of extensor pollicis longus muscle	Ch. 10
dorsal venous arch	Ch. 12
tendon of extensor digiti minimi	Ch. 10
tendon of extensor digitorum muscle	Ch. 10
metacarpophalangeal (met-ah-kar-po-fah-LAN-je-al) joints	
interphalangeal (in-ter-fah-LAN-je-al) joints	

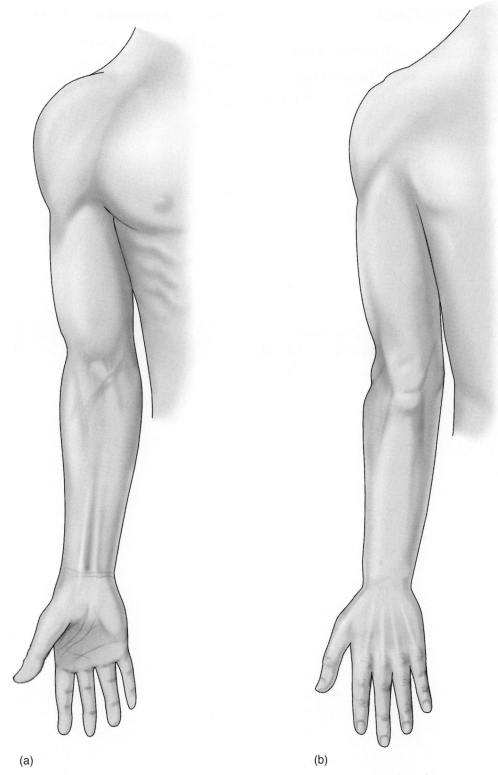

(a)   (b)

FIG. 21.7 Superior appendage, anterior view (a) and posterior view (b).

INFERIOR APPENDAGE

Inferior Appendage, Anterior View

Locate and *label* the following in Figure 21.8(a).

thigh	Ch. 1
rectus femoris muscle	Ch. 10
lateral and medial condyle of the femur	Ch. 7
patella	Ch. 7
patellar ligament	Ch. 10
leg	Ch. 1
tibial tuberosity	Ch. 7
tibia	Ch. 7
medial malleolus of tibia	Ch. 7
lateral malleolus of fibula	Ch. 7
pes	Ch. 1
tendon of extensor digitorum longus muscle	Ch. 10
tendon of extensor hallicis longus muscle	Ch. 10
metatarsophalangeal (met-ah-tar-so-fah-LAN-je-al) joints	
interphalangeal joints	

Inferior Appendage, Posterior View

Locate and *label* the following in Figure 21.8(b).

thigh	Ch. 1
popliteal (pop-LIT-e-al) fossa	
leg	Ch. 1
gastrocnemius muscle	Ch. 10
soleus muscle	Ch. 10
calcaneal tendon	Ch. 10
pes	Ch. 1
calcaneus	Ch. 7

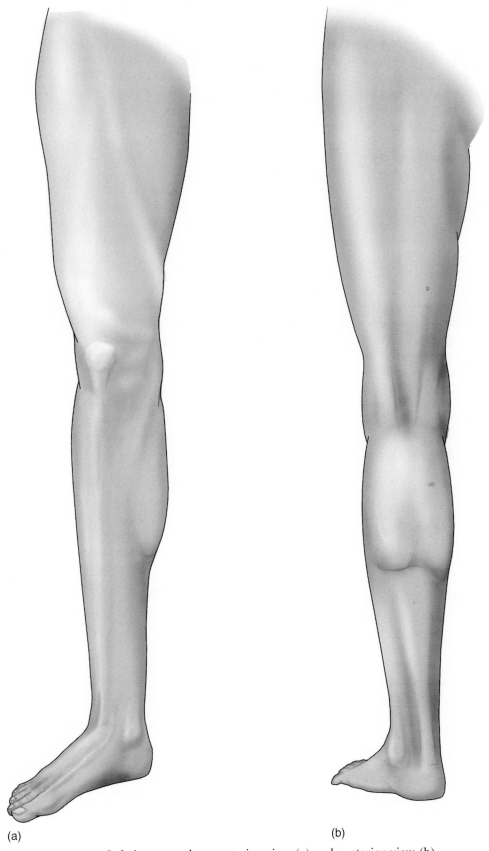

(a)   (b)

FIG. 21.8 Inferior appendage, anterior view (a) and posterior view (b).

SHORT ANSWER

1. *Describe* the significance of the median cubital vein.

2. *Describe* the significance of the iliac crest in giving gluteal injections. Where is it located? (Use the terminology from this chapter.)

3. *Using* the terminology from this chapter, *describe* the location of the "funny bone."

4. If someone takes a blow to the mental region, where was he struck?

5. *Name* the dominant feature of the auricular region.

6. The "six pack" is produced by what structure? Where is it located?

7. If someone touches your lacrimal caruncle, what will be your most immediate response?

8. While kneeling, what structure is contacting the ground?

9. The anatomical term for one's "life line" is _____.

10. *Describe* the location of the following regions of the cranium and face:

temporal region

infraorbital region

mastoid region

buccal region

frontal region

MATCHING

Match the terms in column A with the area or structure of the body on which they appear in column B.

A	B
_____ 1. philtrum	a. abdomen
_____ 2. tragus	b. back
_____ 3. hyoid	c. buttocks
_____ 4. jugular notch	d. chest
_____ 5. linea alba	e. ear
_____ 6. vertebral spines	f. face
_____ 7. gluteal cleft	g. neck
_____ 8. cubital fossa	h. brachium
_____ 9. palmar flexion creases	i. leg
_____ 10 popliteal fossa	j. manus
_____ 11. calcaneal tendon	k. thigh

Glossary

abdominopelvic (ab-dom-ih-no-PEL-vik) Pertaining to the lower abdominal and upper pelvic region.

afferent (AF-er-ent) Moving or conducting toward an area or point.

agonist (AG-ah-nist) The "prime mover" or muscle of reference.

antagonist (an-TAG-ah-nist) The muscle or muscles opposing the agonist.

aponeurosis (ap-o-noo-RO-sis) Dense connective tissue forming a broad tendon.

arteriole (ar-TE-re-ol) A small artery that delivers blood to capillaries.

articulation (ar-tik-yoo-LA-shun) A joint or union of two bones.

aural (AW-ral) Pertaining to the ear or hearing.

brachium (BRA-ke-um) The arm from shoulder to elbow.

buttocks (BUT-oks) Posterior prominence formed by the gluteal muscles.

columnar (ko-LUM-nar) In the form of a column.

coronary (KOR-o-na-re) A condition caused by a decreased blood flow to the heart muscle. Specifically a blocked coronary artery or a branch of a coronary artery. Pertaining to the heart.

costal (KOS-tal) Pertaining to a rib.

cuboidal (kyoo-BOYD-al) In the form of a cube.

deoxygenated (de-OK-si-jen-a-ted) Having no oxygen or having given up oxygen.

depolarization (de-po-lar-ih-ZA-shun) The reversal of a resting potential to form an action potential.

epiphyseal (e-PIF-e-sel) **disk** A cartilaginous "growth line" at the ends of long bones.

exteroception (eks-ter-o-SEP-shun) Perception of information provided by the exteroceptors.

exteroceptor (eks-ter-o-SEP-tor) Sensory nerve terminals stimulated by the external environment.

extrinsic (eks-TRIN-sik) On the outside, related to some other region other than the organ with which it is associated.

facet (FAS-et) A small, facelike articular surface.

fetus (FE-tus) Postembryonic unborn; from the seventh or eighth week of gestation.

fossa (FOS-ah) A depression or trench.

ganglion (GANG-gle-on) A group of nerve cell bodies.

genome (JE-nom) The haploid complement of hereditary factors.

gonad (GO-nad) An ovary or testis.

hallux (HAL-uks) Digit number one of the foot.

haploid (HAP-loyd) A cell such as a gamete with half of the somatic chromosome number (23 in humans).

hematopoiesis (hem-ah-to-poy-E-sis) Blood forming.

hepatic (he-PAT-ik) Pertaining to the liver.

histology (his-TOL-o-je) The study of tissues.

homeostasis (ho-me-o-STA-sis) The maintenance of body metabolism within set boundaries.

hormone (HOR-mon) A chemical messenger secreted by a gland or tissue.

implantation (im-plan-TA-shun) Attachment of the blastocyst to the endometrium.

interosseus (in-ter-OS-e-us) Between bones.

interstitial (in-ter-STISH-al) Between cells of a tissue as in the interstitial cells of the testis that produce testosterone.

intrinsic (in-TRIN-sik)   Found on the inside.

peristalsis (per-ih-STAL-sis)   Movement provided by the smooth muscles to the contents of the digestive tract, ureters, urethra, uterine duct, and vas deferens.

placenta (plah-SEN-tah)   A vascular structure through which the developing embryo and fetus derive nourishment and dispose of wastes in conjunction with the maternal endometrium.

plantar (PLAN-tar)   The sole of the foot.

pollex (POL-eks)   The thumb.

portal (POR-tal)   A pathway or entrance.

protract (pro-TRAKT)   A muscle action that draws an element anteriorly.

pulmonary (PUL-mo-ner-e)   Pertaining to the lungs.

quadriceps (KWOD-ri-ceps)   Having four heads.

repolarization (re-po-lar-ih-ZA-shun)   Return of a neuron to a resting potential.

retroperitoneal (reh-tro-per-ih-to-NE-al)   External to the peritoneum.

simple   Without complexity. In the case of a tissue, without layers.

somatic (so-MAT-ik)   Either pertaining to body cells (as opposed to gametes) or to the body wall (as opposed to the viscera).

spermatogenesis (sper-ma-to-JEN-e-sis)   The testicular process that produces sperm.

sphincter (SFINK-ter)   A ring or rings of smooth muscle that constrict(s) a passage.

spine (spin)   A sharp projection from the body or main part of a bone.

squamous (SKWAH-mus)   Flat or scalelike.

stratified (STRAT-ih-fid)   In layers.

subarachnoid (sub-ah-RAK-noyd)   Deep to the arachnoid membrane of the meninges and superficial to the pia.

synergist (SIN-er-gist)   Acting together.

systemic (sis-TEM-ik)   Pertaining to, or acting on, the body as a whole.

thermoregulation (therm-mo-reg-yoo-LA-shun)   Control of body temperature.

umbilical (um-BIL-ih-kal)   Pertaining to the umbilicus or navel.

ventral (VEN-tral)   The belly or toward the belly.

venule (VEN-yool)   Small vessels that join to form veins.

visceral (VIS-er-al)   The organs of the thoracic, abdominal, and pelvic cavities.

Photo Credits

Figures CDA-2 to CDA-11: Stephen Lewin. Figures CDA-14 to CDA-16: Courtesy Mark Nielsen. Figure HA-1: Biophoto Associates/Photo Researchers. Figures HA-3 to HA-17: Courtesy Victor Eroschenko. Figure HA-18: ©CABISCO/Visuals Unlimited. Figure HA-19: Astrid & Hanns-Frieder Michler/Photo Researchers. Figure HA-20: Lester Bergman/Project Masters, Inc. Figure HA-21: Courtesy Victor Eroschenko. Figure HA-22: Ed Reschke. Figures HA-23 and HA-24: Courtesy Victor Eroschenko. Figure HA-25: ©CABISCO/Visuals Unlimited. Figure HA-26: Courtesy Victor Eroschenko. Figures HA-27 and HA-28: Biophoto Associates/Photo Researchers. Figure HA-29: Ed Reschke. Figure HA-30: Lester Bergman/Project Masters, Inc. Figures HA-31 and HA-32: Courtesy Victor Eroschenko.

Index

G

galea aponeurotica, 104
gallbladder, 277, 283
gastroesophageal, 280, 284
genome, 260
germinal, 219, 295
gingivae, 278
gland, 259
    adenoid, 219, 266
    adrenal, 260, 261
    bulbourethral, 297
    ceruminous, 34
    gastric, 192, 200, 215, 216, 280
    hypophyseal, 46, 60
    hypophysis, 196, 229, 230, 236, 260, 261
    hypothalamus, 227, 229, 230, 236, 252, 260, 261
    juxtaglomerular, 289
    mammary, 35, 218
    mucous, 270, 280
    pancreas, 201, 216, 260, 261, 282, 283, 284
    parathyroid, 260, 262
    parotid, 279
    pineal, 229, 236, 260, 262
    pituitary, 46, 229
    prostate, 297
    salivary, 277, 279
    sebaceous, 34
    serous, 172
    sublingual, 279
    sudoriferous, 34
    thymus, 206, 211, 217, 219, 260, 262
    thyroid, 108, 109, 186, 260, 261, 262, 268, 270, 308
glans, 297
glenoid, 76, 112, 113
glomerular, 289
gluteal tuberosity, 83
golgi, 18
gonad, 290
granulosum, 34
groove, 46, 69, 77, 111

H

hallux, 7, 140, 144, 146
haploid, 290, 295
helix, 308
hepatic, 192, 193, 194, 200, 211, 212, 214, 215, 216, 283, 284
histology, 23, 44, 277
homeostasis, 259
hormone, 172, 205, 259, 260, 261, 277
humor, 252
hymen, 293

I

ileocecal, 280
ileocolic, 192, 193, 284
ileum, 216, 280, 284
illuminator, 16
implantation, 290
incisive, 57
infraorbital, 57, 306
infundibulum, 229, 230, 235, 293
integument, 33
internus, 130
interventricular, 176, 178, 180, 181
intervertebral, 65, 89
intestine, 193, 201, 277, 280, 281, 282, 284
iris, 252, 254, 307

J

jejunal, 192, 193
jejunum, 216, 280, 284
joint
    amphiarthroses, 87, 89
    articular, 45, 65, 66, 67, 68, 83, 84, 89
    articulate, 46
    articulation, 46
    diarthrosis, 87, 89
    frontozygomatic, 57
    interphalangeal, 312, 314
    knee, 7
    lamdoidal, 58, 62
    metacarpophalangeal, 120, 312
    metatarsophalangeal, 314
    sacroiliac, 80
    suture, 57, 58, 59, 89
    symphasis, 80, 89, 134, 157, 310
    synarthrosis, 87, 89
    synchondrosis, 89
    syndesmosis, 89
    synovial, 87, 89

K

kidney, 202, 287

L

labia, 104, 293, 294
lacerum foramen, 59, 60
lacrimal foramen, 57
lacteal, 218, 277, 281
lacunae, 30, 45
lamellae, 44, 45
lamina, 65, 66, 67
laryngopharynx, 265, 266
larynx, 108, 109, 265, 266, 268, 269, 270, 308

THIS PAGE FOR YOUR NOTES AND DRAWINGS

THIS PAGE FOR YOUR NOTES AND DRAWINGS

THIS PAGE FOR YOUR NOTES AND DRAWINGS

THIS PAGE FOR YOUR NOTES AND DRAWINGS

THIS PAGE FOR YOUR NOTES AND DRAWINGS

THIS PAGE FOR YOUR NOTES AND DRAWINGS

HISTOLOGY ATLAS

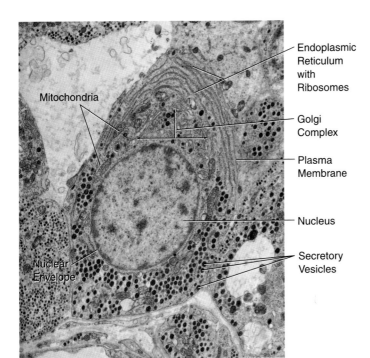

FIGURE HA-1 Animal Cell **TEM** 11875x

Labels: Endoplasmic Reticulum with Ribosomes, Mitochondria, Golgi Complex, Plasma Membrane, Nucleus, Secretory Vesicles, Nuclear Envelope

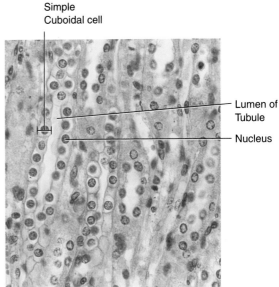

FIGURE HA-3 Simple Cuboidal Epithelium (Kidney Tubules) **LM** 117x

Labels: Simple Cuboidal cell, Lumen of Tubule, Nucleus

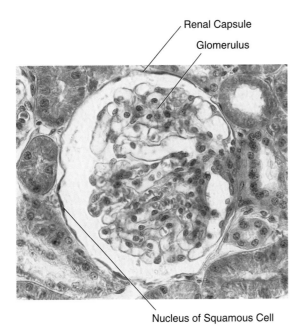

FIGURE HA-2 Simple Squamous Epithelium (Renal Corpuscle) **LM** 117x

Labels: Renal Capsule, Glomerulus, Nucleus of Squamous Cell

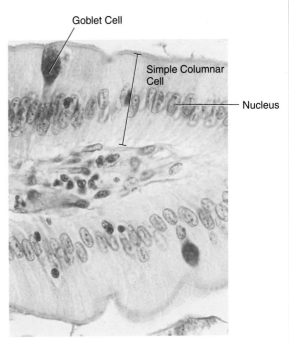

FIGURE HA-4 Simple Columnar Epithelium **LM** 117x

Labels: Goblet Cell, Simple Columnar Cell, Nucleus

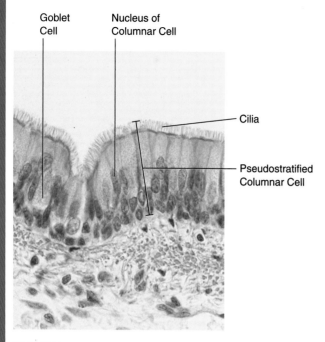

Goblet
Cell

Nucleus of
Columnar Cell

Cilia

Pseudostratified
Columnar Cell

FIGURE HA-5 Ciliated, Pseudostratified Columnar
Epithelium (Trachea) **LM** 341x

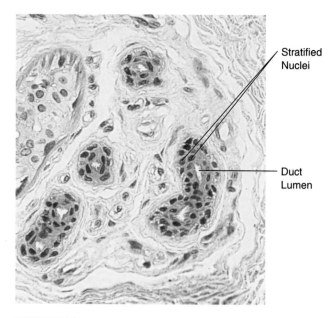

Stratified
Nuclei

Duct
Lumen

FIGURE HA-7 Stratified Cuboidal Epithelium (Sweat
Gland Duct) **LM** 403x

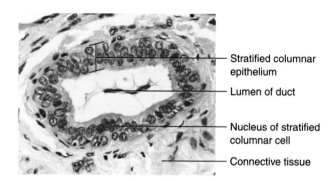

Stratified columnar
epithelium

Lumen of duct

Nucleus of stratified
columnar cell

Connective tissue

FIGURE HA-8 Stratified Columnar Epithelium (Salivary
Gland Duct) **LM** 420x

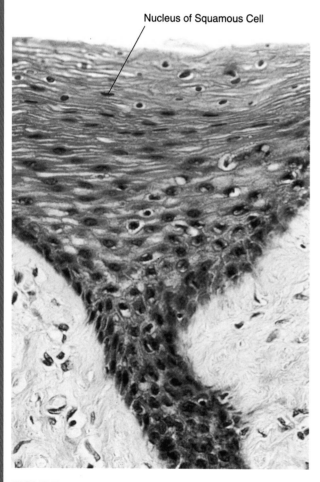

Nucleus of Squamous Cell

FIGURE HA-6 Nonkeratinized Stratified Squamous
Epithelium (Vagina) **LM** 50x

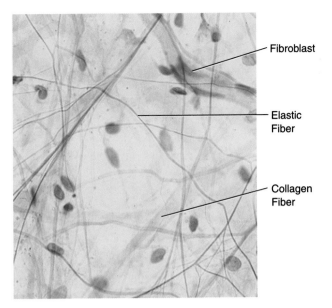

FIGURE HA-9 Areolar Connective Tissue (Mesentary) **LM** 537x

Fibroblast

Elastic Fiber

Collagen Fiber

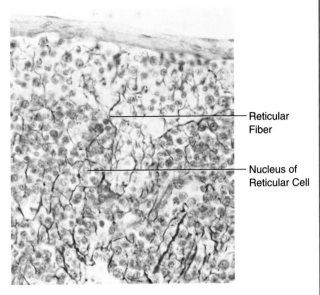

FIGURE HA-11 Reticular Connective Tissue (Lymph Node) **LM** 422x

Reticular Fiber

Nucleus of Reticular Cell

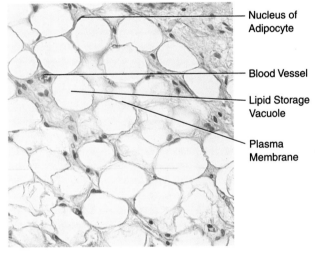

FIGURE HA-10 Adipose Tissue **LM** 386x

Nucleus of Adipocyte

Blood Vessel

Lipid Storage Vacuole

Plasma Membrane

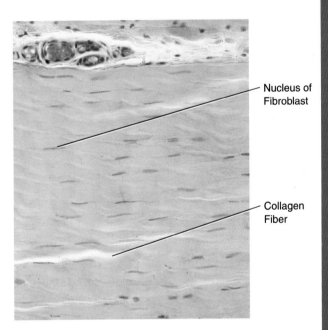

FIGURE HA-12 Dense Regular Connective Tissue (Tendon) **LM** 453x

Nucleus of Fibroblast

Collagen Fiber

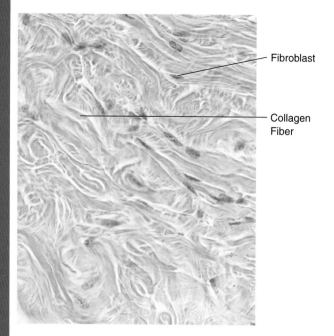

FIGURE HA-13 Dense Irregular Connective Tissue (Dermis) **LM** 294x

Fibroblast

Collagen Fiber

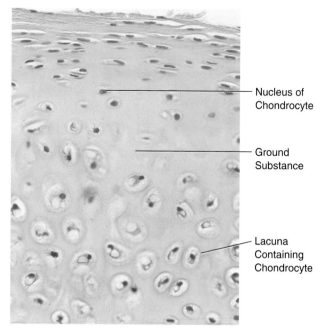

FIGURE HA-15 Hyaline Cartilage (Trachea) **LM** 295x

Nucleus of Chondrocyte

Ground Substance

Lacuna Containing Chondrocyte

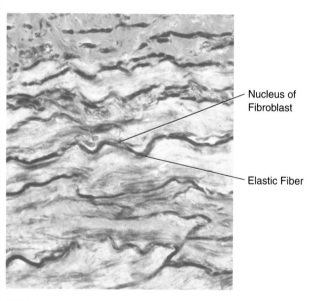

FIGURE HA-14 Elastic Connective Tissue (Aorta) **LM** 448x

Nucleus of Fibroblast

Elastic Fiber

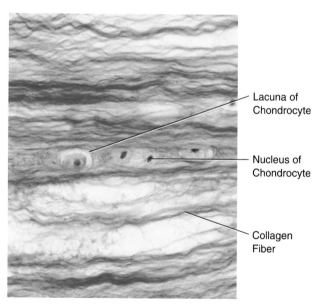

FIGURE HA-16 Fibrocartilage (Intervertebral Disk) **LM** 434x

Lacuna of Chondrocyte

Nucleus of Chondrocyte

Collagen Fiber

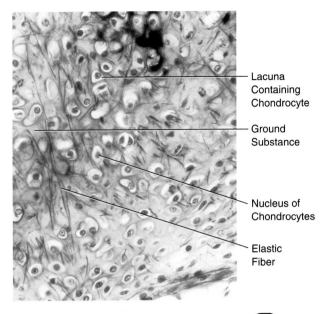

FIGURE HA-17 Elastic Cartilage (Epiglottis) **LM** 272x

Labels: Lacuna Containing Chondrocyte; Ground Substance; Nucleus of Chondrocytes; Elastic Fiber

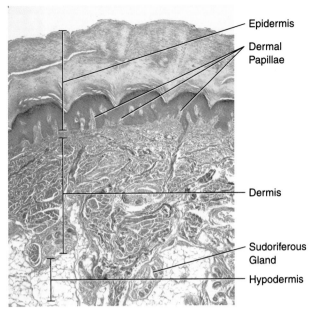

FIGURE HA-19 Human Skin, Hairless **LM** 30x

Labels: Epidermis; Dermal Papillae; Dermis; Sudoriferous Gland; Hypodermis

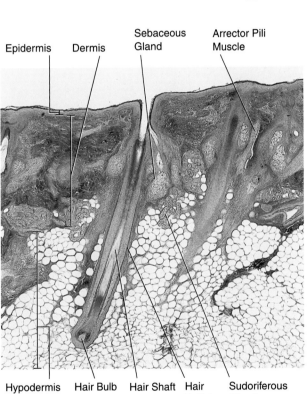

FIGURE HA-18 Human Skin, Hairy **LM** 19.5x

Labels: Epidermis; Dermis; Sebaceous Gland; Arrector Pili Muscle; Hypodermis; Hair Bulb; Hair Shaft; Hair Follicle; Sudoriferous Gland

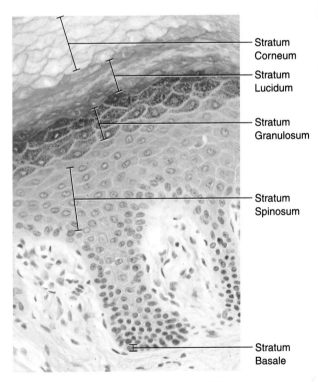

FIGURE HA-20 Human Epidermis **LM** 442x

Labels: Stratum Corneum; Stratum Lucidum; Stratum Granulosum; Stratum Spinosum; Stratum Basale

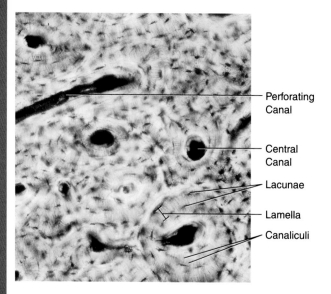

FIGURE HA-21 Compact Bone **LM** 92x

Perforating Canal

Central Canal

Lacunae

Lamella

Canaliculi

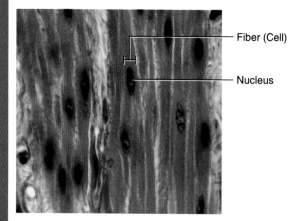

FIGURE HA-22 Visceral Muscle **LM** 840x

Fiber (Cell)

Nucleus

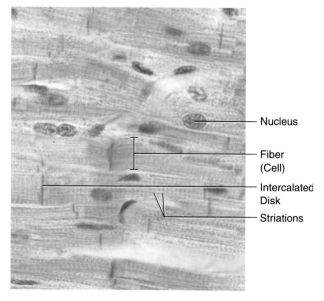

FIGURE HA-23 Cardiac Muscle **LM** 688x

Nucleus

Fiber (Cell)

Intercalated Disk

Striations

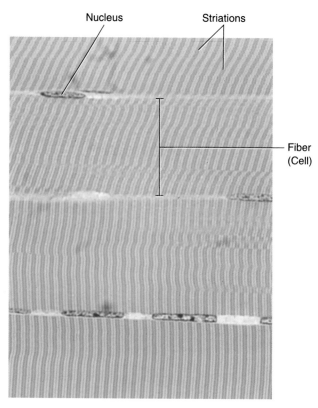

Nucleus   Striations

Fiber (Cell)

FIGURE HA-24 Skeletal Muscle **LM** 672x

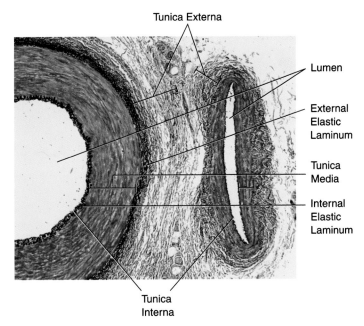

FIGURE HA-25 Artery (left) and Vein (right), cross sections **LM** 40x

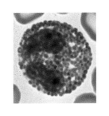

Eosinophil **LM** 1500x

FIGURE HA-27

Basophil **LM** 1500x

FIGURE HA-28

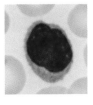

Large lymphocyte
LM 1500x

FIGURE HA-29

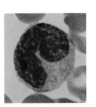

Monocyte **LM** 1500x

FIGURE HA-30

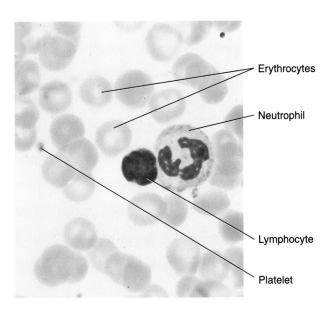

Erythrocytes

Neutrophil

Lymphocyte

Platelet

FIGURE HA-26 Human Blood **LM** 1017x

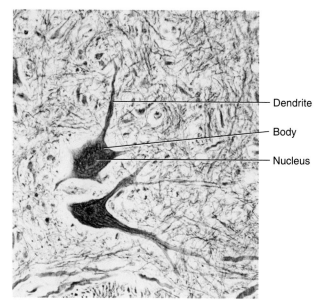

FIGURE HA-31 Multipolar Neuron (Spinal Cord)
LM 694x

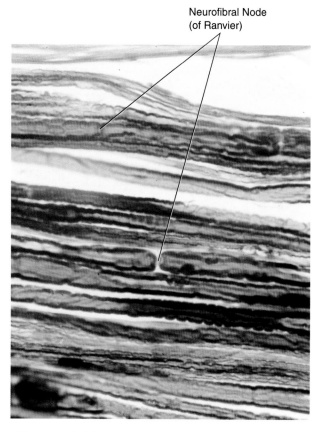

FIGURE HA-32 Nodes of Ranvier (Nerve, Long. Sec.)
LM 905x

CAT DISSECTION ATLAS

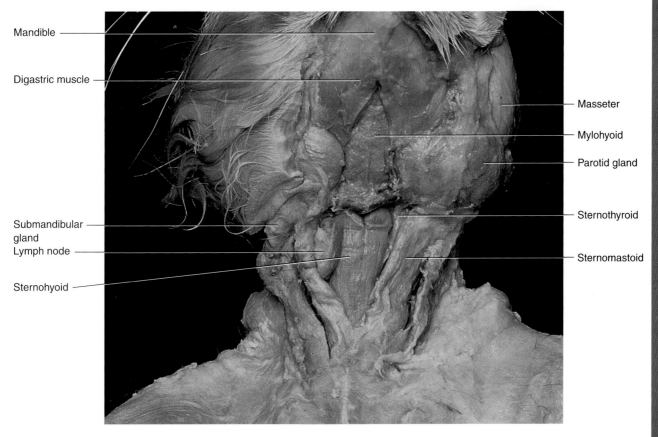

Mandible

Digastric muscle

Masseter

Mylohyoid

Parotid gland

Submandibular gland

Lymph node

Sternohyoid

Sternothyroid

Sternomastoid

CDA-1 Superficial muscles of the head and neck, ventral view.

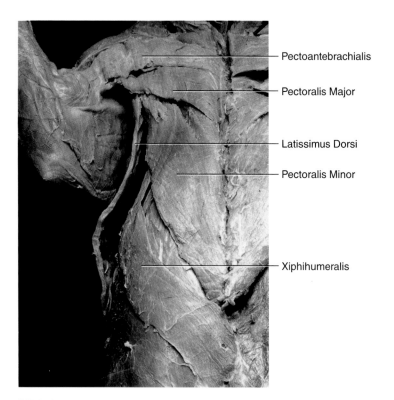

- Pectoantebrachialis
- Pectoralis Major
- Latissimus Dorsi
- Pectoralis Minor
- Xiphihumeralis

CDA-2 Superficial muscles acting on the scapula and brachium, ventral view.

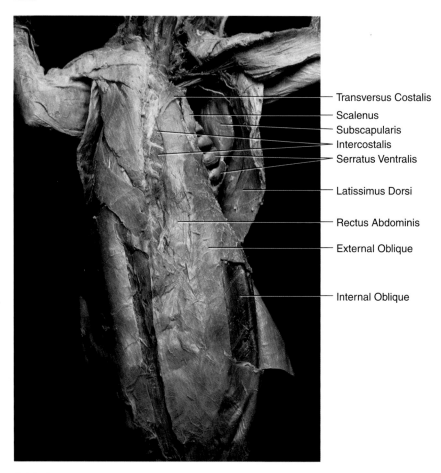

- Transversus Costalis
- Scalenus
- Subscapularis
- Intercostalis
- Serratus Ventralis
- Latissimus Dorsi
- Rectus Abdominis
- External Oblique
- Internal Oblique

CDA-3 Deep muscles acting on the scapula and brachium and superficial and deep muscles of the thorax and abdomen, ventral view.

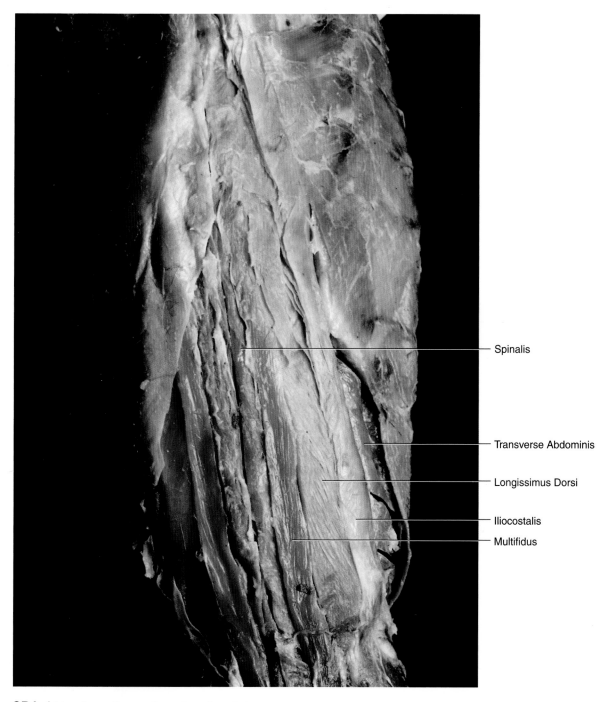

CDA-4 Muscles acting on the spine, dorsal view.

Spinalis

Transverse Abdominis

Longissimus Dorsi

Iliocostalis

Multifidus

CDA-5 Superficial muscles acting on the scapula and brachium, lateral view.

Spinotrapezius
Clavotrapezius
Acromiotrapezius
Latissimus Dorsi
Spinodeltoid
Acromiodeltoid
Clavodeltoid

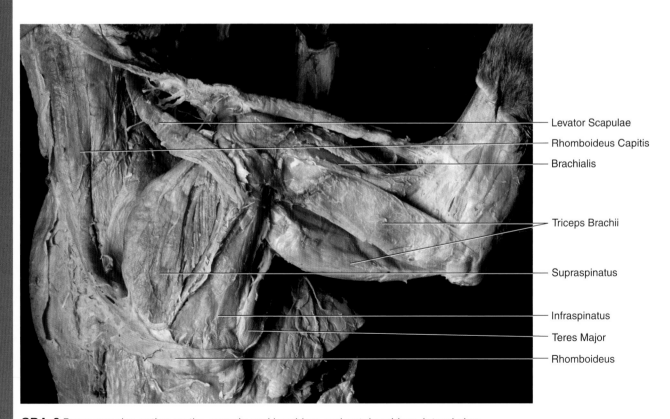

CDA-6 Deep muscles acting on the scapula and brachium and antebrachium, lateral view.

Levator Scapulae
Rhomboideus Capitis
Brachialis
Triceps Brachii
Supraspinatus
Infraspinatus
Teres Major
Rhomboideus

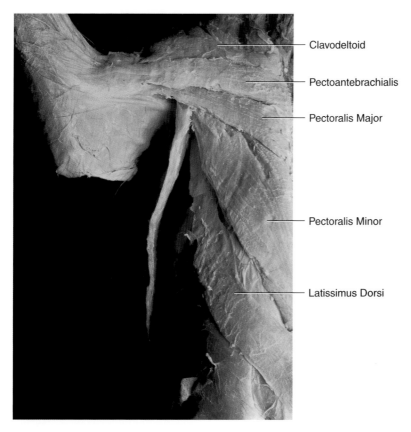

CDA-7a Superficial muscles acting on the brachium and antebrachium, medial view.

Clavodeltoid

Pectoantebrachialis

Pectoralis Major

Pectoralis Minor

Latissimus Dorsi

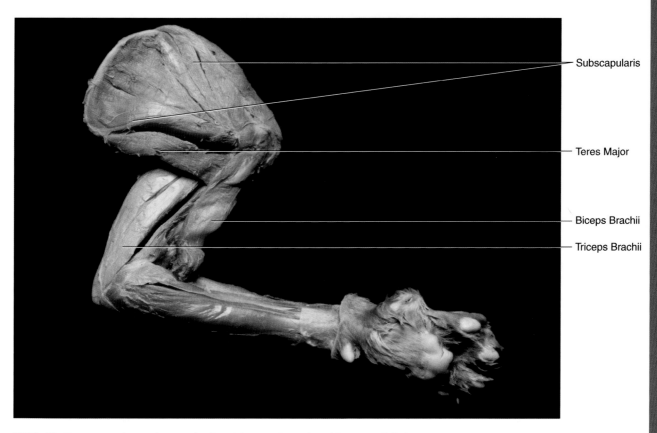

Subscapularis

Teres Major

Biceps Brachii

Triceps Brachii

CDA-7b Deep muscles acting on the brachium and antebrachium, medial view.

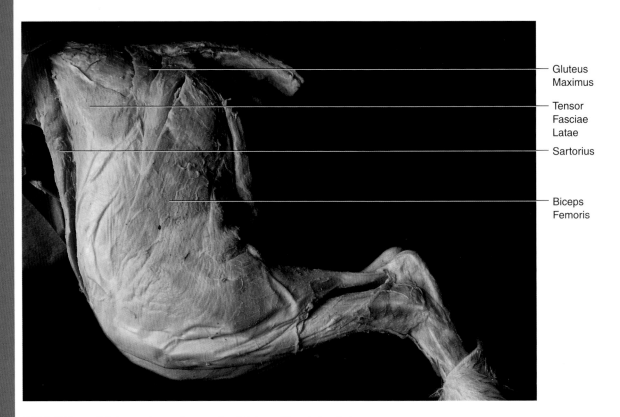

Gluteus Maximus

Tensor Fasciae Latae

Sartorius

Biceps Femoris

CDA-8 Superficial muscles acting on the thigh and leg, lateral view.

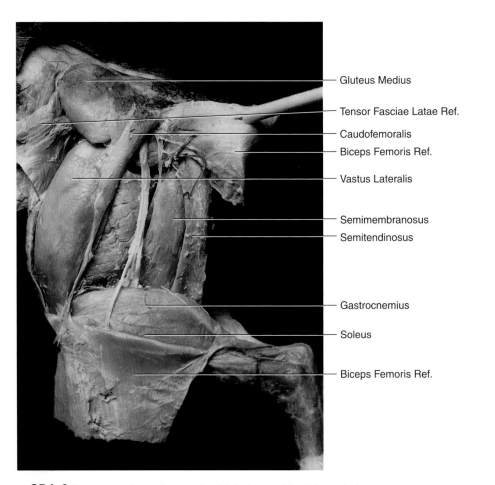

Gluteus Medius

Tensor Fasciae Latae Ref.

Caudofemoralis

Biceps Femoris Ref.

Vastus Lateralis

Semimembranosus

Semitendinosus

Gastrocnemius

Soleus

Biceps Femoris Ref.

CDA-9 Deep muscles acting on the thigh, leg and foot, lateral view.

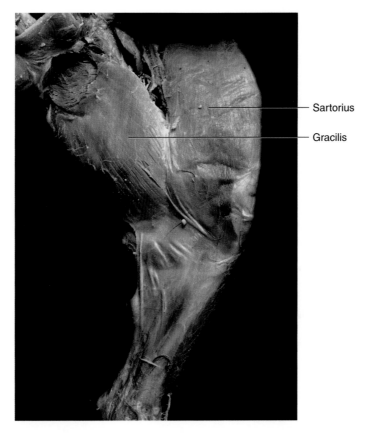

CDA-10 Superficial muscles acting on the thigh and leg, medial view.

— Sartorius

— Gracilis

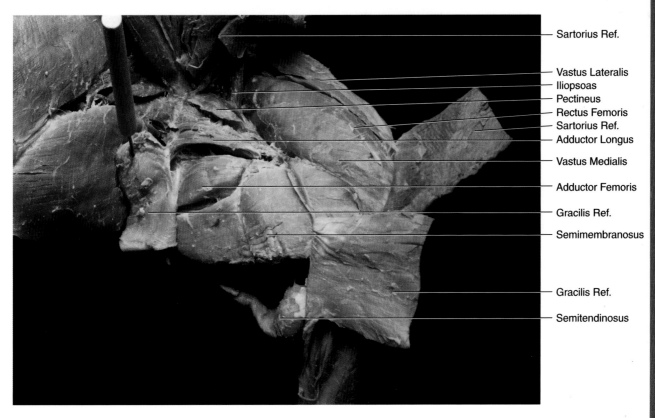

CDA-11 Deep muscles acting on the thigh and leg, medial view.

— Sartorius Ref.

— Vastus Lateralis
— Iliopsoas
— Pectineus
— Rectus Femoris
— Sartorius Ref.
— Adductor Longus

— Vastus Medialis

— Adductor Femoris

— Gracilis Ref.

— Semimembranosus

— Gracilis Ref.

— Semitendinosus

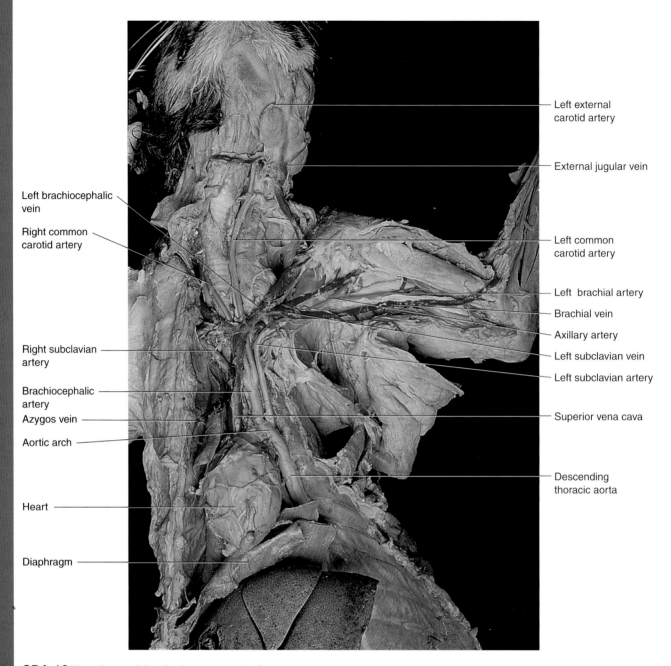

Left external
carotid artery

External jugular vein

Left brachiocephalic
vein

Right common
carotid artery

Left common
carotid artery

Left brachial artery

Brachial vein

Axillary artery

Right subclavian
artery

Left subclavian vein

Left subclavian artery

Brachiocephalic
artery

Azygos vein

Superior vena cava

Aortic arch

Descending
thoracic aorta

Heart

Diaphragm

CDA-12 Vessels cranial to the heart, ventral view.

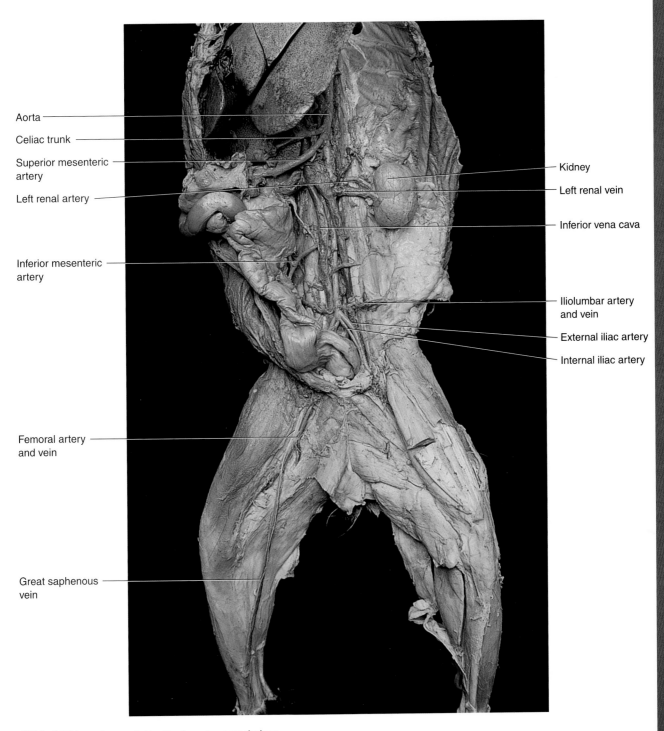

Aorta

Celiac trunk

Superior mesenteric
artery

Left renal artery

Inferior mesenteric
artery

Femoral artery
and vein

Great saphenous
vein

Kidney

Left renal vein

Inferior vena cava

Iliolumbar artery
and vein

External iliac artery

Internal iliac artery

CDA-13 Vessels caudal to the heart, ventral view.

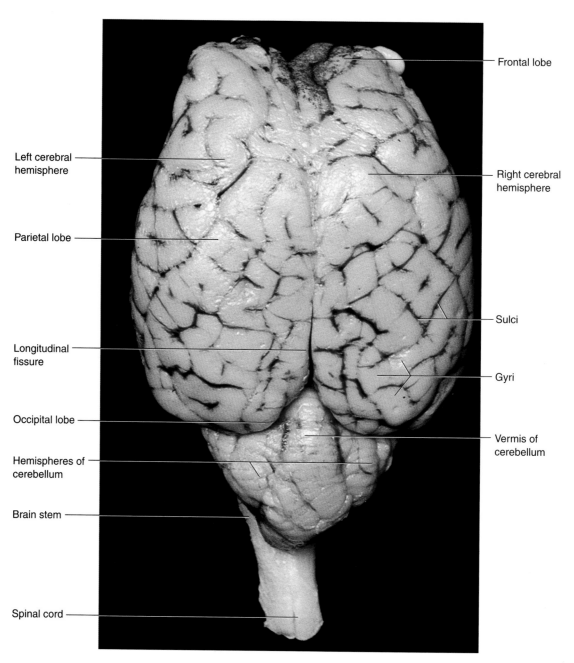

Frontal lobe

Left cerebral
hemisphere

Right cerebral
hemisphere

Parietal lobe

Sulci

Longitudinal
fissure

Gyri

Occipital lobe

Vermis of
cerebellum

Hemispheres of
cerebellum

Brain stem

Spinal cord

CDA-14 Sheep brain, dorsal view.

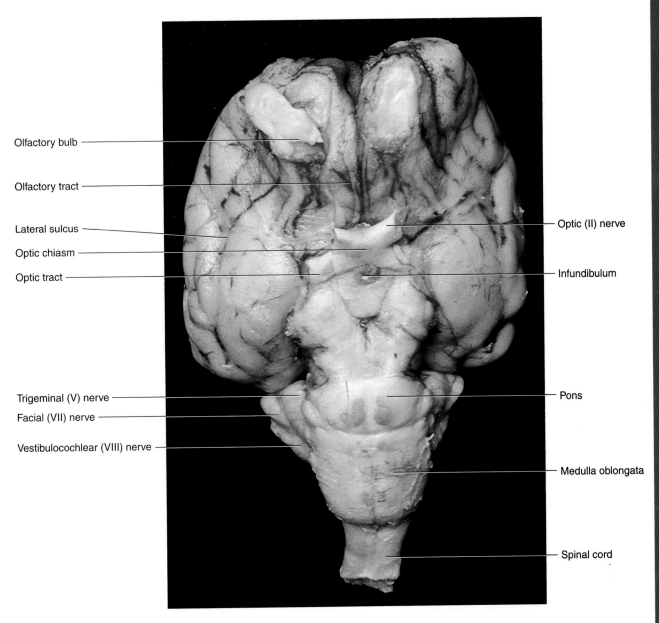

Olfactory bulb

Olfactory tract

Lateral sulcus

Optic chiasm

Optic tract

Optic (II) nerve

Infundibulum

Trigeminal (V) nerve

Facial (VII) nerve

Vestibulocochlear (VIII) nerve

Pons

Medulla oblongata

Spinal cord

CDA-15 Sheep brain, ventral view.

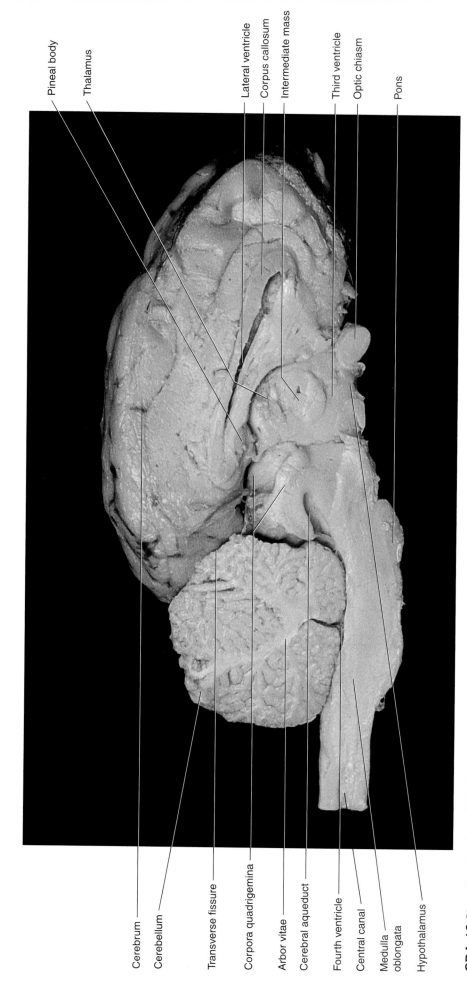

Pineal body
Thalamus
Lateral ventricle
Corpus callosum
Intermediate mass
Third ventricle
Optic chiasm
Pons

Cerebrum
Cerebellum
Transverse fissure
Corpora quadrigemina
Arbor vitae
Cerebral aqueduct
Fourth ventricle
Central canal
Medulla oblongata
Hypothalamus

CDA-16 Sheep brain, midsagittal view.

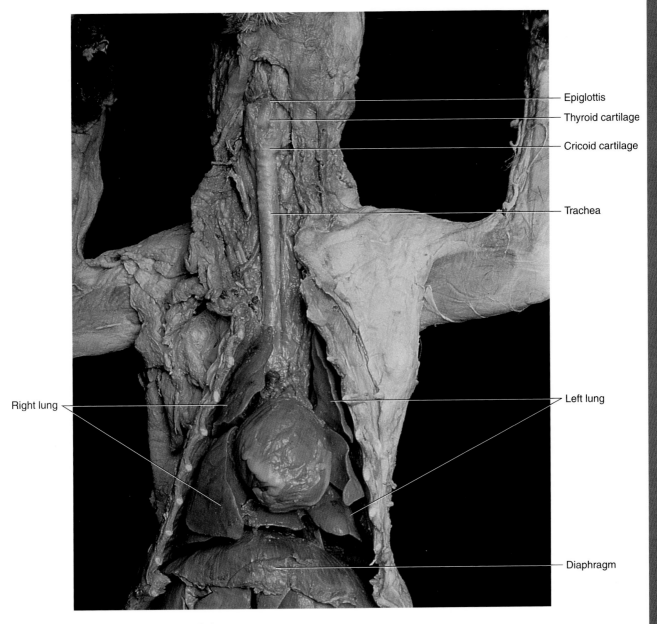

Epiglottis

Thyroid cartilage

Cricoid cartilage

Trachea

Right lung

Left lung

Diaphragm

CDA-17 Respiratory system, ventral view.

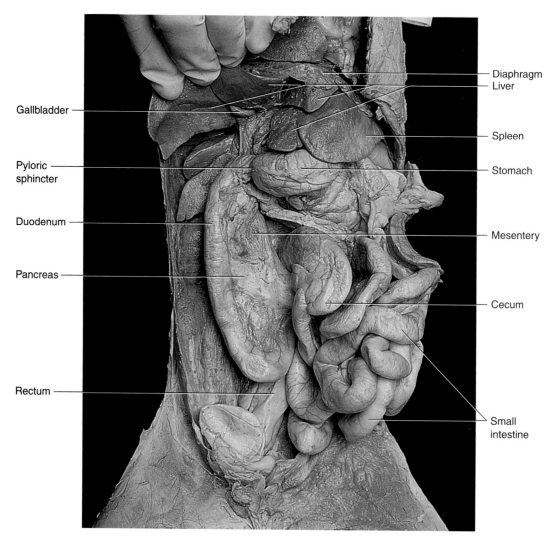

CDA-18 Digestive system, ventral view.

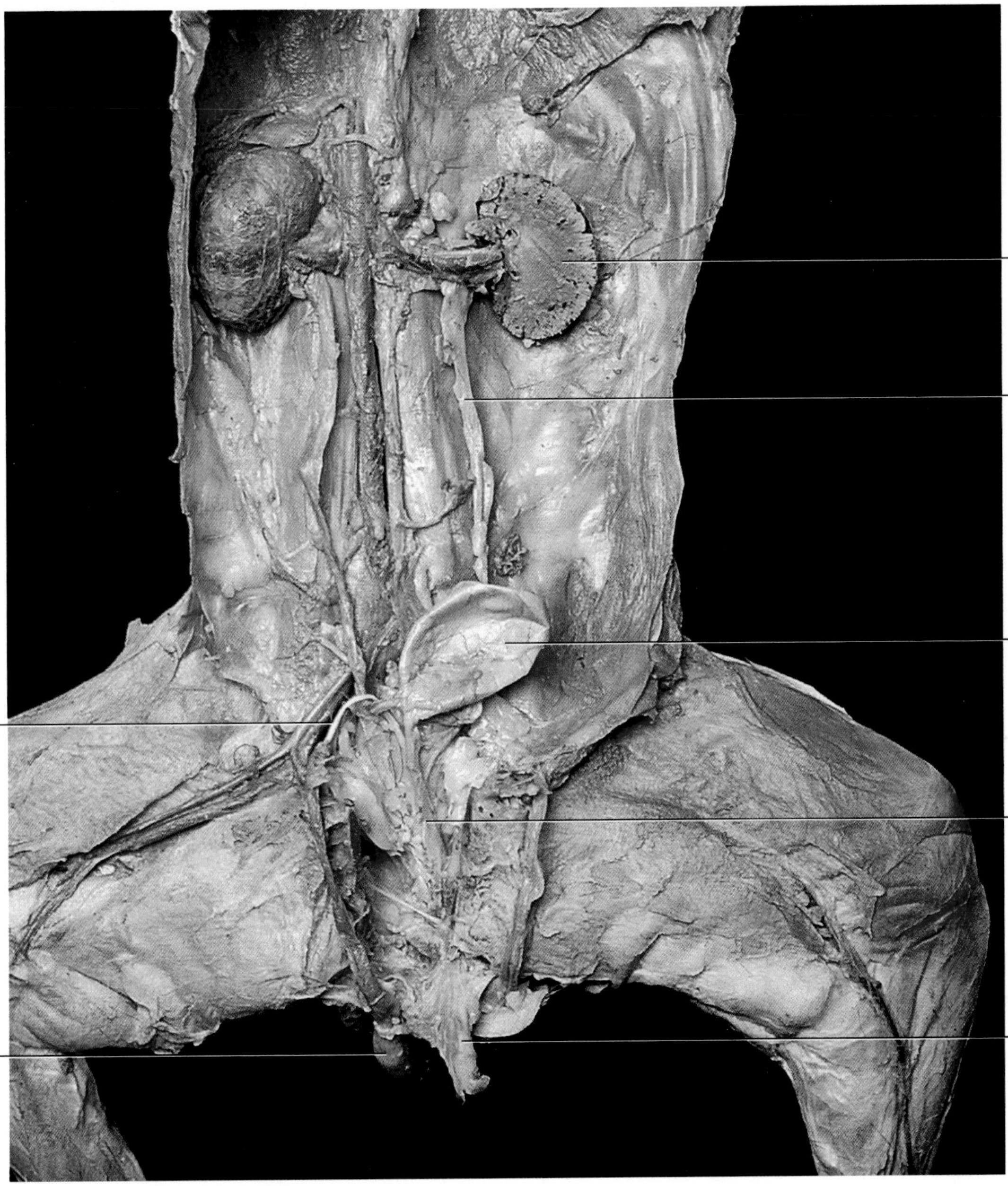

Left kidney
(sectioned)
Left ureter
Urinary
bladder
Ductus
deferens
Urethra
Penis
Testis

CDA-19 Male urogenital, ventral view.

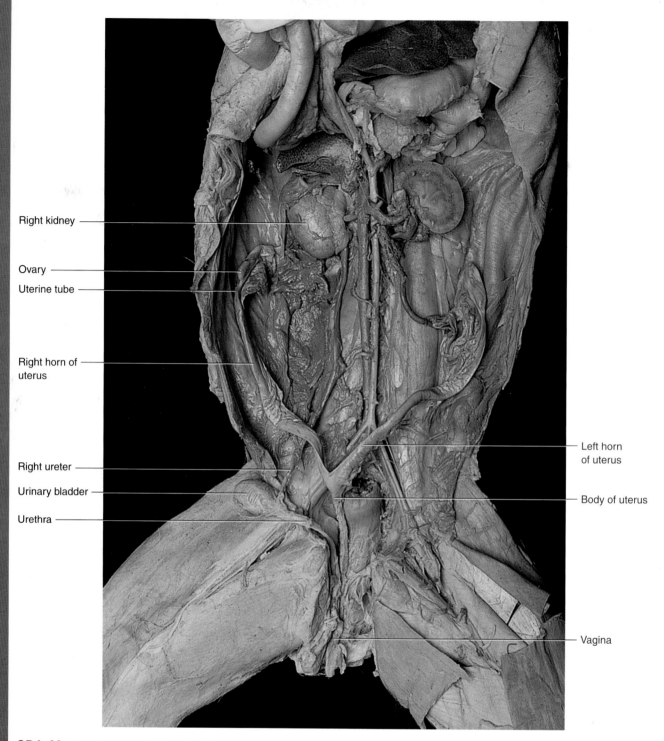

Right kidney

Ovary

Uterine tube

Right horn of
uterus

Right ureter

Urinary bladder

Urethra

Left horn
of uterus

Body of uterus

Vagina

CDA-20 Female urogenital, ventral view.